Linking Flexibility, Uncertainty and Variability in Manufacturing Systems

Managing un-planned change in the automative industry

HENRIQUE LUIZ CORRÊA
University of São Paulo and
Warwick Business School

Avebury
Aldershot · Brookfield USA · Hong Kong · Singapore · Sydney

Published by
Avebury
Ashgate Publishing Limited
Gower House
Croft Road
Aldershot
Hants GU11 3HR
England

Ashgate Publishing Company
Old Post Road
Brookfield
Vermont 05036
USA

British Library Cataloguing in Publication Data

Corrêa, Henrique Luiz
Linking Flexibility, Uncertainty and Variability in Manufacturing Systems: Managing Un-planned Change in the Automative Industry
I. Title
338.4767

ISBN 1 85628 620 7

Printed and Bound in Great Britain by
Athenaeum Press Ltd, Newcastle upon Tyne.

Contents

Preface

This is a research book. It shows some results of four years of research work performed during a Ph.D. study leave from the University of São Paulo, spent at two English institutions: Brunel University and The University of Warwick. During this period, I was able to visit a number of English and Brazilian manufacturing companies, most of them belonging to the automotive industry, and study them with respect to the way their managers deal with the issue of manufacturing flexibility. While learning from these companies and departing from the already advanced work of a number of other researchers in the field, I also developed some thoughts about this fascinating subject.

During the last 15 years, flexibility has become one of the hottest buzz words in the world of manufacturing. After the quality wave, flexibility now seems to be one of the most desired characteristics of the manufacturing companies. It is frequently said that flexibility can help shorten manufacturing lead-times, reduce batch sizes and inventory levels, introduce new products more frequently and quicker; flexibility can help support market segmentation policies, product customization. Of course, everybody wants to be flexible!... But what is it to be flexible? How do we know whether flexibility can really help the company to compete or whether it is just another fad? What are the strategic implications of flexibility? Are there types and dimensions of flexibility? Do we need different types of flexibility in different situations? What is it that makes the difference for a manufacturing resource to be considered more or less flexible? Is it possible to measure flexibility?

This book is an attempt to answer some of these questions. It treats flexibility at the manufacturing systems level, not the individual resources' level. The underlying assumption behind this book is that flexibility is only justified if it can actually help increase the organization's competitive power.

Chapter 1 discusses the concept of flexibility in the context of the manufacturing strategies.

Chapter 2 is a literature review of the field of manufacturing flexibility at the manufacturing systems level.

Chapter 3 discusses two of the categories considered by many authors as reasons for system managers to be interested in developing flexibility: environmental uncertainty and variability of outputs.

Chapter 4 is a theory building exercise. Based on the concepts discussed in the first three chapters and on the case studies, an analytical model is developed which tries to link the notions of flexibility, uncertainty and variability. The point of contact between the three concepts is found to be the 'management of unplanned change', a principle explained throughout Chapter 4.

Chapter 5 compares the empirical and analytical results of the research with the current literature, identifies some avenues for further research and critically reviews the research done.

The book was organized in order that a reader interested only in the concepts and results of the research can read the chapters of the main body of the book without having to go through the details regarding methodology issues and the case studies themselves. However, it is also possible that the readers who are interested in analyzing the methodological aspects of the researchcan do so by consulting the Appendixes.

Appendix 1 describes the methodology used in the research and Appendix 2 describes the in depth case studies performed.

Henrique Luiz Corrêa
University of São Paulo

Acknowledgements

My special thanks to people and institutions who were of great help throughout this research project and also during the write up of the book:

My research supervisor during my stay in England, Professor Nigel Slack, who was always supportive and who knew how to challenge me at the right moments. His good humoured constant encouragement has helped me renew my interest in an academic career. In a way, the ideas developed in this book are only an extension of the original work on flexibility that Nigel has developed during the last decade. His work was both inspiring and influential in this research.

Professors Ray Wild, from Henley The Management College and Richard Ormerod from Warwick Business School, for the valuable comments on the thesis which originated this book.

Mrs. Josephine Gooderham, Avebury Commissioning Editor, for the opportunity to have this work published.

My parents, Szonya and Alberto Corrêa, who fully supported me not only throughout these five years of research but at all moments of my life. To them, all my love and gratefulness.

Maria Teresa Corrêa de Oliveira, for the support during the writing-up process and for the patience in helping review the manuscript.

Dr. Barbara Hemais and Dr. Carlos Alberto Hemais for their detailed review of the manuscript.

Mr. Guido Casanova for his help with the preparation of the manuscript.

The Brazilian People, for sponsoring me during the whole project, through a scholarship given by CNPq, "Conselho Nacional de Desenvolvimento Científico e Tecnológico" and also through a study leave, given by EPUSP, "Escola Politécnica da Universidade de São Paulo".

The Staff at the School of Industrial and Business Studies of the University of Warwick, as well as my colleagues who were also research students at

SIBS, where I found the ideal conditions and motivation for the development of this project in its most important stages.

The Staff of the Department of Manufacturing and Engineering Systems at Brunel University for the support during the first stages of my research.

The five Brazilian and three British companies which agreed to host me for numerous days, in particular the managers who spent long hours of their valuable time with me in formal and informal meetings and plant tours, kindly answering my "endless" questions.

My colleagues at the Department of Production Engineering of the University of São Paulo, especially Mr. Irineu Gianesi, Mr. Mauro Caon, Professor Afonso Fleury, Dr. Guilherme Plonsky, Dr. Mario Salerno, for the interesting comments on parts of the manuscript.

Dr. Peter McKiernan, for the encouragement and the valuable comments on various parts of the manuscript.

My dear friends, Dr. Ana Julia Jatar de Haussmann and Dr. Ricardo Haussmann, who taught me so much during our stay at SIBS.

The British People who hosted me in such a warm way for three and a half unforgettable years.

The managers who were my students in courses on Manufacturing Strategy and Manufacturing Flexibility at the "Warwick Manufacturing Group - University of Warwick", United Kingdom, at the "Instituto de Estudios Superiores de Administración", Venezuela, and at the "Fundação Carlos Alberto Vanzolini - USP", Brazil, with whom I discussed some of the ideas contained in this book and who gave me valuable insights.

1 Flexibility in the context of manufacturing strategy

Introduction and objective

The objective of this Chapter is to place manufacturing flexibility in its strategic context. The first part of the Chapter discusses why manufacturing strategy has been one of the most studied issues of recent years in the operations management literature. The second part discusses the views found in the literature about the manufacturing strategy contents - objectives and decision areas - and process. Manufacturing flexibility is described, in the third part, as one of the competitive criteria which the organization may pursue in order to enhance its competitiveness. The role of manufacturing flexibility as a first and as a second order competitive criterion is analyzed, in terms of enhancing the organization's competitiveness. A second order competitive criterion is one which is not a competitive criterion in itself, but an indirect criterion which influences the performance of the organization in terms of other criteria e.g. costs and delivery speed.

Manufacturing strategy has increasingly been regarded by academics and practitioners as having an important contribution to make to enhanced competitiveness. The growth of the literature in manufacturing strategy has matched the growth of interest in the area. Within the literature three main reasons are identified for its newly found importance.

The first of these is the increased pressure owing to the growing international manufacturing competitiveness. The second is the increased potential to be gained from the development of new manufacturing technologies and the third is the development of a better understanding of the strategic role of manufacturing. Each of these stimuli will be examined in turn.

The changing international manufacturing competition

During the last 30 years the relative competitive positions occupied by the leading industrial countries have changed substantially. Some traditional industrial nations have been outperformed by other countries, of which Japan is the most evident example. The United States and the United Kingdom have had their leading positions challenged and in many cases lost them, e.g. in the automobile market, long dominated by American companies (Womack et al., 1990; Hill, 1993).

Buffa (1984), considering the Japanese manufacturing industry, notices that the industries in which they have excelled - motor cycles, domestic appliances, automobiles, cameras, hi-fi, and steel production - had existing, already developed markets with established market leaders. According to the same author, Japanese companies may have succeeded, partially because of their Finance and Marketing related skills, but largely because of the high quality and low cost which they achieved through a sharp manufacturing practice which most of the Western manufacturers may not have been able to match. Buffa (1984) observes that the Japanese companies were using the improvements which they had been achieving in Manufacturing as their main competitive advantage, as opposed to the Western companies, which had considered Manufacturing as a 'solved problem', focusing their attention on getting competitive advantage through achieving excellence in marketing their products and managing their financial issues.

Not only are Japanese companies on average more cost efficient than most Western companies (though there are many exceptions of Western companies which have maintained or improved their competitive position in the world market during the last decades), but they are competing and winning based also on their better quality and reliability performance as well as on their better responsiveness to the market needs and opportunities. In the introduction of new products, for instance, Japanese car manufacturers cut their product development times (the period between the earliest stages of design and the manufacture of a new model) to an average of less than four years compared to six to eight years in Europe and America.

The reasons behind the changes

There is, in general, agreement that (initially, at least) Western companies lacked an effective response to the Japanese challenge. The reasons behind this lack of an effective response by most of the Western companies which faced such a challenge are various, according to the literature. Hayes and Wheelwright (1984) summarize some of them in five main points:

Financial considerations The assessment of companies and their manager's performance based predominantly on short term considerations may have induced managers to avoid long term investments which might have resulted in a more effective manufacturing. Managers may not have decided to invest in improvements whose results would only show in the long term because they needed short term performance. Kaplan (1984) argues that the traditional accounting methods, developed basically to support mass production, undermine today's production, because of their short termism and inadequacy to support production in the new competitive reality. The new competitive environment requires broader product ranges, faster and more frequent product introductions and products of higher quality levels.

Technological considerations Western managers would have been less sophisticated, imaginative and even interested in dealing with technological considerations than the overseas competitors, focusing attention predominantly on financial and marketing issues.

Excessive specialization and/or lack of proper integration Western managers would have tended to separate complicated issues into simpler, specialized ones to a greater degree than their foreign counterparts without having developed proper integration to pull the differentiated responsibilities together and to be able to deal with the total picture.

Lack of focus The separating and specializing mentality would have led many Western firms to diversify away from their core technologies and markets. They would have tended to adopt the *portfolio* approach, used by stocks and bonds investors. This approach considers that diversifying is the best way to hedge against random set-backs. Manufacturing, however, would not be subject only to random set-backs but, more significantly, to carefully orchestrated attacks from competitors who focus all their resources and energy on one particular set of activities. Focused manufacturing is based on the idea that simplicity, repetition, experience and homogeneity in tasks breed competence (Skinner, 1974).

Inertia Skinner (1985) observes that most factories in the Western world were not managed very differently in the 1970s from the way they were in the 1940s or 1950s. Such practices, continues Skinner, may have been adequate when production management issues centered largely on efficiency and productivity. However, the problems of operations managers moved far beyond mere physical efficiency. On top of this, managers considered that the production problems were solved, directing attention and resources toward other issues such as distribution, packaging and advertising. According to Hill

(1985), there has been a failure, conscious or otherwise, of Western industries and the society at large to recognize the size of the foreign competitive challenge, its impact on their way of life, and consequently to recognize the need for change.

The result of the concurrence of the five factors above is that Western plants and equipment were allowed to age. What one day had been technological advantage eroded by the decline in expenditure and attention to issues such as new products research and development and new process technologies (Hayes and Wheelwright, 1984). Then, Hayes and Wheelwright conclude, 'in the beginning of the 1970s, US companies found themselves pitted against companies that did compete on dimensions such as defect-free products, process innovation and delivery dependability. Increasingly, they found themselves displaced first in international markets and then in their home market as well'. This conclusion can also be extended to many non-US Western companies.

One Western country which represents an exception in the loss of competitiveness to Japanese companies is West Germany. In spite of the upward pressure on the Deutsche Mark during most of the 1970s and the economic stagnation and political turmoil the Germans have faced in recent years, the German economy remains strong in its most important area, the manufacturing sector. German manufactured exports grew vigorously during the 1970s and Germany's productivity growth rate in the manufacturing sector actually increased in the 1970s, whereas it dropped in other Western industrial countries. The good German performance is based on factors such as technological strength throughout the company's managerial hierarchy, intense product and customer orientation, orientation toward the growth and stability of the firm, among other manufacturing related factors.

When we compare German and Japanese approaches to manufacturing management, some points in common can be found - such as the technological strength throughout the managerial hierarchy - but quite different characteristics can also be noted, for instance the strong emphasis on the team approach and participative management style, a norm in Japanese firms and less emphasized in German ones. However in the literature the authors recognize that Japan and West Germany managed to find their competitive way. They were better able to exploit their strengths and mitigate their weaknesses. They showed that there were alternative paradigms to the one in vogue since World War II. This does not mean that their practices are unequivocally good or necessarily superior to practices in other countries. The main point is that their superior performance called the attention of Western managers and academics to the need to rethink their own manufacturing practices in order to find their own effective way, one which is adequate to their needs and potential and appropriate to the new reality of the world market.

The development of new manufacturing technologies

Manufacturing Technology is regarded as one of the most important decision areas within the manufacturing management function. Traditionally, manufacturing management has influenced manufacturing technology to a much greater extent than the other way round.

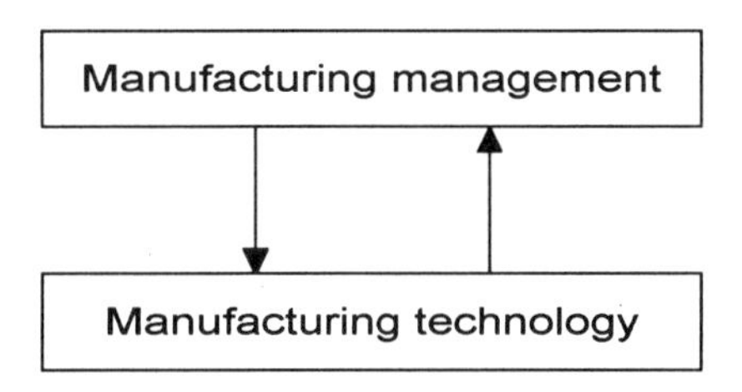

Changes in the manufacturing technology were for a long time slow and gradual not calling for profound changes in its management methods and techniques. With the new micro-electronics and information handling technology being incorporated into the process technologies, the resulting changes were not gradual and did not follow the usual pattern. A new paradigm was established. Computer controlled flexible machines challenged the once well established concept of *economies of scale* because they have the potential of making changeover times negligible. The concept of *economies of scope*[1] (Goldhar and Jelinek, 1983) started to gain importance.

The new flexible technology made it possible to produce different products at the same rates which had only been possible with mass production, with single or a few products. The strict one-to-one relationship between product and process life cycles (Hayes and Wheelwright, 1984)[2] would not apply any more (Stecke and Raman, 1986).

According to Voss (1986) the development of new process technologies has been of such proportion that it has outstripped the ability of people to use it at its full advantage. The potential capabilities of the new technologies include reducing design-to-production lead times, reducing order to delivery lead times, improving the conformance quality of products, among others. This can change the way organizations compete in the market place. New manufacturing technologies start influencing the manufacturing management more relevantly. Questions such as: 'how can the new technologies make us more competitive?' and 'how can the new technologies change the way we compete?' become the key questions. The new technological paradigm has called for a new management approach.

Robotics, CAM (computer aided manufacturing), FMS (flexible manufacturing systems), among other newly available manufacturing process technology labels, are now current terms in the manufacturing environment and they came to challenge some previously well established concepts, e.g. the economies of scale. However, despite the optimism of some authors, there is evidence in the literature that the new process technologies have not proved to be as influential as initially thought. The expectations and also the investments with regard to the new technologies were initially high but the results, although considerable, have not followed suit.

The choice of the adequate process technology is more than ever a critical strategic decision. Each choice of process will bring with it strategic implications for a business in terms of: response to the market needs, manufacturing capabilities and characteristics, level of investment required, unit costs involved, type of control and style of management. Traditionally, manufacturing technology has been seen in relatively narrow terms. Specialist engineers have tended to think in technical rather than operating or strategic terms. The availability of the new manufacturing technologies, according to Hayes and Wheelwright (1984), calls for at least three kinds of strategic fit: the first is internal to the manufacturing function and relates to coordinating technology activities with operating policies and systems. The second has to do with the consistency between internal and external activities or, in other words, with meshing manufacturing capabilities with the capabilities and needs of the other functions and the firm's overall competitive strategy. The third fit relates to the consistency over time, ensuring that the firm's technology evolves in a directed fashion so that as technological capabilities are renewed and augmented, they reinforce and expand the firm's competitive position. A comprehensive and strategic perspective is thus more than ever necessary to deal with the new manufacturing technologies to ensure an adequate choice and appropriate management so that they actually contribute at their full potential to the business competitiveness.

In summary, without a clear strategic direction with regard to manufacturing, the new manufacturing technologies can become an expensive 'solution in search of a problem'. In this sense, one of the aims of manufacturing strategy is to give the organization strategic direction with regard to manufacturing issues, technology included, making sure that not only the technologies but also the people and the infrastructure used are consistent with the strategic objectives of the business.

The development of a better understanding of the strategic role of manufacturing

Skinner is one of the authors who first recognized and called attention to the strategic role of the manufacturing function in business and corporate competitiveness. In his early articles (Skinner, 1969, 1974) the author states that the acceleration of foreign competition, technological changes in production and information handling equipment and the social changes in the work force call for profound changes in manufacturing function management. The potential of manufacturing as a competitive weapon and the concept of using manufacturing as a strategic asset could no longer be overlooked by managers who should abandon a number of old assumptions about manufacturing. A new approach would be necessary, in order to respond to the new reality. Skinner makes a number of points, which can be grouped under four main headings:

i) Manufacturing can be a formidable competitive weapon. Manufacturing matters have been a missing link in corporate strategy and companies which intend to be competitive should start to consider manufacturing in a strategic way.

ii) Cost efficiency is not the only contribution which manufacturing can provide to business competitiveness. The assumption that the main criteria for evaluating factory performance are efficiency and cost should be challenged and new criteria should be adopted which evaluate how the firm is competing rather than how efficient it is.

iii) Trade-offs must be made and priorities established between manufacturing performance criteria. According to this view, a good factory could not simultaneously excel in all performance criteria such as low cost, high quality, minimum investment, short cycle times and rapid introduction of new products.

iv) Competitive manufacturing must be focused. Companies should focus each plant on a concise and manageable set of products, technologies, volumes and markets and develop manufacturing policies and supporting services so that they focus on one explicit manufacturing task instead of many inconsistent and conflicting ones.

Since the seminal work of Skinner, a number of authors have addressed the strategic role of the manufacturing function. Hayes and Whellwright (1984) called attention to the need to transform the manufacturing role from being primarily reactive to being *proactive*, where the manufacturing function contributes actively to the achievement of competitive advantage.

Another point which is made by some authors, e.g. Slack (1990) refers to the fact that the complexity of the manufacturing function calls for strategic management. According to Slack, manufacturing is almost certainly the largest (both in terms of people and capital employed), probably the most complex and arguably the most difficult of all the functions within the organization to manage. Manufacturing strategy would thus involve developing a manufacturing system or a set of manufacturing resources which enable the organization to compete more effectively in the market place.

Hill (1993) argues that the need for a manufacturing strategy to be developed and shared by the business has to do not only with the critical nature of manufacturing within the corporate strategy but also with a realization that many of the decisions in manufacturing are structural in nature. Therefore, unless the issues and consequences are fully appreciated by the business, then it can be locked into a number of manufacturing decisions which may take years to change. Changing them is costly and time consuming, but even more significantly, the changes will possibly come too late.

Focused manufacturing: an increasingly important concept

Although the manufacturing function is regarded as one of the most complex to manage within the organization, what creates the complexity is not the technology dimension but the number of aspects and issues involved, the inter related nature of these and the level of fit between the manufacturing task and its internal capability (Hill, 1993). The level of complexity involved depends largely on corporate and marketing strategy decisions, made within the business, where the competitive priorities are established. These competitive priorities are established because a manufacturing system cannot excel in all aspects of performance at the same time. Trade-offs must be made. Different types of performance demand different manufacturing resources organized in different ways (Slack, 1989a). An organization which competes predominantly on cost efficiency, for instance, by manufacturing in high volumes, would need different resources (possibly more dedicated machines) in order to compete effectively if compared to an organization competing on product customization, making products to order (which would possibly call for more general purpose flexible equipment).

This is the rationale behind the concept of focused manufacturing. According to this view, for the effective support of competitive business strategy the manufacturing function should focus each part of its manufacturing system on a restricted and manageable set of products, technologies, volumes and markets so as to limit the manufacturing objectives in which it is trying to excel. This means that if an organization has different products or product groups competing in different ways, then its manufacturing function should reflect this in the way it is subdivided so as to maintain focus on what is most important for its competitiveness in the market place.

If a company competes on a broad range of products, the decision to adopt the concept of focused manufacturing can have the disturbing implication of calling for major investments in new plants and new equipment to break down the existing complexity. One alternative approach which helps to avoid major investments is a solution that does not involve selling big multipurpose facilities and decentralizing them into small ones. The solution could be the more practical approach of the 'plant-within-a-plant', where the existing facility is divided both organizationally and physically into plants within the original plant. Each of them would have its own facilities. Each plant-within-the-plant can this way concentrate on its particular manufacturing task, using its own work force management approaches, production control systems, organizational structure and so forth. Each plant-within-the-plant would quickly gain experience by focusing and concentrating every element of its work on those limited essential objectives which constitute its manufacturing task or focus.

The idea of focus should thus permeate all the process of formulation and execution of the business and manufacturing strategies. The establishment of competitive priorities and the decision making process should also take the idea of focus into consideration, in order to make sure that the manufacturing function can really excel in what it is expected to.

Although it is intuitive and appealing, having gained broad support lately among academics and practitioners, the concept of focused manufacturing still lacks further empirical support[3]. Further research is needed to test its assumptions and prescriptions.

The manufacturing strategy process

In general, authors agree on the prime aim of manufacturing strategy which, according to them, is to support the organization's achievement of a long term sustained competitive advantage. It is also clear that for most of the authors the development of a manufacturing strategy should follow a top-down approach. Hayes and Wheelwright (1984), Skinner (1985), Slack (1991), Gregory and Platts (1990) and Hill (1993) suggest hierarchical models in which corporate strategy drives business strategies. This in turn drives the strategies of manufacturing and other functional areas within the business unit. Although the dominant approach for formulating a manufacturing strategy is top-down it seems that as long as the manufacturing function reaches more developed stages in Hayes and Wheelwright's (1984) four-stage classification[4], the capabilities developed by the manufacturing function start influencing more strongly and to a certain extent also start driving the corporate and business strategies, with a somewhat bottom-up view being aggregated to the dominant top-down approach. Summarizing, the authors seem to agree with regard to the basic top-down, break down approach for manufacturing strategy formulation where corporate strategy drives business strategy, which in turn drives the strategies of manufacturing and other function areas within the business unit (Marketing and R&D, among others), breaking down strategies of one level into objectives of the following level and so on up to the level of the manufacturing strategy decision areas, defined in the next section.

The manufacturing strategy contents

The manufacturing strategy contents are divided into objectives and decision areas. Each will be discussed in turn.

Manufacturing strategy objectives

The principal aim of manufacturing strategy is to support the organization's achievement of a long term sustained competitive advantage. Competitive advantage is achieved through manufacturing by managing the organization's resources in order to provide an appropriate mix of performance characteristics or competitive priorities. In general terms, the authors seem to agree about the main objectives which the manufacturing systems should pursue, although the terminology they use varies widely, making accurate comparisons difficult.

Skinner (1985) and Fine and Hax (1985) define the manufacturing objectives as having four broad dimensions: cost, quality, delivery and flexibility.

Wild (1980) divides the objectives into two groups: the ones related to customer service and the ones related to resource productivity. Resource productivity refers to how efficiently the manufacturing resources are utilized. In terms of customer service, three main competitive factors are identified: product specification (design and performance levels), cost (price and expense levels) and timing (delivery time). Wild suggests that apart from the level achieved, another dimension should be considered for each factor: reliability, which would be the principal direct contribution of operations management.

Slack (1983) extends Wild's analysis and adds volume and mix of output to the factors and flexibility to the dimensions. Volume relates to the ability to manufacture at a particular rate, mix relates to the ability of manufacturing products in a particular mix and flexibility relates to how far and how easily a system could change what it is doing. In a later work, Slack (1991) summarizes his objectives in five dimensions: quality, cost, responsiveness, dependability and flexibility.

Buffa (1984) finds that organizations use four manufacturing related dimensions in order to compete: cost, quality of products and services, dependability of supply (delivery dependability) and flexibility/service which include the ability to accommodate variations in the product or service, availability of spare parts, field services, among others.

Hayes and Wheelwright (1984) identify five main competitive priorities: low cost/price, high performance (product features, tolerances and customer services), dependability (product, delivery and field service), flexibility (broad line, customized products, fast response and delivery time) and innovativeness (new products, latest technology).

Hill (1985) introduces the concept of *order winning criteria* which are the objectives manufacturing should pursue in order to win orders in the market place. The main order winning criteria are, according to Hill, price, quality, delivery speed, delivery reliability, product and color range and design leadership.

A composite view of the literature results in the following main competitive priorities (with terminology adaptations):

i) Cost: manufacturing and distribution of the products at low costs;

ii) Cost dependability: meeting required or intended costs;

iii) Productivity: achievement of a better utilization of process technology, labor and material resources;

v) Product quality: manufacturing of products with high performance and conformance to standards;

vi) Range of products: manufacturing a broad range of products;

vii) Innovativeness: introduction of new products or processes;

viii)Delivery speed: reacting quickly to customer orders;

ix) Delivery dependability: meeting delivery schedules or promises; and

x) Flexibility: changing easily what is being done.

Figure 1.1 summarizes the competitive priorities of selected authors. A brief analysis of Figure 1.1 shows that there are basically four competitive priorities which are explicitly present in all the authors' lists: cost efficiency, product quality, delivery speed and delivery dependability. Flexibility is another competitive factor which is present in most of the authors' lists.

year	Skinner (1978)	Wild (1980)	Buffa (1984)	Hill (1985)	Fine &Hax (1985)	Hayes et al. (1988)	Slack (1991)
Cost	x	x	x	x	x	x	
Cost dependability		x					
Productivity		x					x
Product quality	x	x	x	x	x	x	x
Range of products				x		x	
Innovativeness						x	
Delivery speed	x*	x		x	x	x	x
Delivery dependability	x*	x	x	x	x	x	x
Flexibility	x		x		x	x	x

Figure 1.1 Manufacturing competitive priorities according to some selected authors
* Skinner mentions only delivery

Although the authors mention sets of objectives which the organizations should pursue in order to achieve long term competitive advantage, the relative importance given to them varies, according to the particular market in which the organizations compete. There are, nevertheless, some objectives which, at a certain point in time, gain special and generalized attention.

In the mid 1970s, special emphasis on the objective of quality arose, mainly as a response to the increasing competitive power of some Japanese companies which managed to show the inappropriateness of the assumption that the concepts of quality and cost efficiency are incompatible: they were producing better products at lower costs. A vast literature then emerged, diffusing techniques such as statistical process control, zero-defect campaigns, quality control circles, among others. The whole world became mobilized, seeking quality improvements.

From the late 1970s on, in addition to the objective of quality, flexibility became a major concern of managers and academics.

De Meyer (1986), reporting research with large manufacturing companies in Europe, Japan and the United States, suggests that while 1975-85 could well be labeled as an era where manufacturers discovered that there was no trade-off to be made between quality of product and service and efficiency of the production system, 1985-95 had the potential of becoming an era where manufacturers would discover that flexibility is not necessarily contradictory with the pursuit of cost efficiency. De Meyer argues that Japanese companies are leading other countries in this tendency and justifies such leadership by arguing that these companies' current performance in terms of quality gives them sufficient lead over American and European competitors to concentrate their efforts on the trade-offs between flexibility and cost.

Stalk and Hout (1990) suggest that time will be the next source of competitive advantage. According to this view, the companies which manage to reduce the time span of their processes will take the lead in the near future. Since flexible systems tend to respond quicker to market needs, it seems that flexibility and time-based competitiveness are somehow linked as manufacturing objectives. This point will be further discussed later in this Chapter.

Manufacturing strategy decision areas

Hayes and Wheelwright (1984) and Skinner (1985) characterize manufacturing strategy as a consistent pattern of many individual decisions that affect the ability of the firm to achieve long term sustained competitive advantage. Because the manufacturing function is complex, these authors and others have categorized the individual decisions in strategic decision areas[5] and provide a framework to analyze and shape a pattern for the decisions which should be consistent with the organization's objectives. The problem of lack of standard terminology makes it difficult to compare the various categories proposed by the authors. However, it is possible to see some level of agreement amongst them with regard to the decision areas, as can be seen in Figure 1.2 .

A composite view of the manufacturing strategy decision areas in the relevant literature, with the adjustment of terminology, converges at the following 10 main decision areas:

i) Capacity: amount, type, timing, responsiveness;

ii) Facilities: layout, size, location, specialization, maintenance policies;

iii) Technology: equipment, automation, linkages, capability, flexibility;

iv) Vertical integration: direction, extent;

v) Work force: skill levels, wage policies, employment security;

vi) Quality: defect prevention, monitoring, intervention, standards;

vii) Material flow: sourcing policies, decision rules, role of inventories, responsiveness;

viii)New products: focus, responsiveness, frequency;

ix) Performance measurement: priorities, standards, methods; and

x) Organization: centralization, leadership style, communication, decision making.

Wild (1980) - design and specification of the process and systems, location, layout, capacity and capability, design of work and jobs, scheduling of activities, quality, inventory, maintenance, replacement of facilities, and performance measurement.

Buffa (1984) - capacity/location, product/process technology, workforce and job design, operating decisions, supplier and vertical integration, and positioning of system.

Skinner (1985) - plant and equipment, production planning and control, organization and management, labor and staffing, and product design and engineering.

Hill (1985) - choice of alternative processes, trade-offs embodied in the process choice, role of inventories in the process configuration, function support, manufacturing systems, control and procedures, work structuring, and organizational structure.

Fine and Hax (1986) - capacity, facilities, vertical integration, process/technology, scope and new products policy, human resources, quality management, manufacture infrastructure, and vendors relations.

Hayes et al. (1988) - capacity, facilities, technology, vertical integration, work force, quality, production planning and control, new product development, performance measurement, and organization.

Slack (1989a) - design of the manufacturing system, management of product response, management of materials flow, long term capacity, management of demand response, and manufacturing control system.

Figure 1.2 Decision areas in manufacturing strategy according to some selected authors

Manufacturing strategy - conclusions

Considerable progress has been made in the definition of manufacturing strategy since Skinner's early conceptual work. The most appropriate way to define manufacturing strategy currently seems to be a composite view of some of the researchers in the field.

Manufacturing strategy can be defined as a framework whose central task is to enhance long term sustained competitiveness by organizing manufacturing resources and shaping manufacturing related decisions so as to provide an appropriate mix of desired performance characteristics.

It seems that a general agreement exists in the scarce literature about the process of developing manufacturing strategy. Leong et al. (1990) notice that process research has been relatively neglected conceptually and almost totally neglected empirically by the literature. The proposed models seem to be reasonable, but empirical research work is still needed in order to validate them. Another important question is whether the models are appropriate for the current and future competitive environment. Many authors, for instance, agree upon a top-down approach for the manufacturing strategy formulation process. At the same time, they also agree that it is desirable that the manufacturing function has a proactive role rather than only reactive within the organization. This suggests that a bottom-up component should be incorporated to the dominant top-down approach. However, apart from some overlooked feed-back loops, the bottom-up component is not explicitly present in any of the frameworks found in the literature. Most of the frameworks also suggest that the manufacturing strategy replanning process should be triggered by time, with a replanning period which generally varies from six months to one year. However it seems that with the increasing turbulence of the markets, the frameworks should also consider a formal means for the replanning process to be triggered by relevant events whenever they happen. One year or even six months appears to be too long a time for a company to wait to redirect its strategy in case some relevant aspect of the environment changes substantially. The competitors, subject to the same changes, can develop more responsive systems and therefore react to them more quickly, which can be an increasingly important advantage in times of time-based competitiveness (see Corrêa and Gianesi, 1992).

In terms of the contents of manufacturing strategy, which are by far more fully explored in the literature than the process, there seems to be agreement, to a certain extent, among the authors, about the approach they adopt. Most of them divide the overall problem into two main content areas: the set of manufacturing strategic objectives, which are sometimes called competitive priorities (Leong et al., 1990), or order winning criteria (Hill, 1985) and the decision areas. Leong et al. notice that empirical work on the contents of manufacturing strategy has been produced more substantially than on process. However, such empirical work, according to the authors, would tend to be predominantly descriptive. They suggest that the time has come to move on to testing bigger ideas and building new theory (such as the manufacturing focus concept, broadly discussed and reasonably accepted but still lacking empirical evidence).

The literature points out five main manufacturing objectives: cost, quality, delivery speed, delivery dependability and flexibility. In historical terms, some general trends of these objectives can be observed. In the 1940s and 1950s, cost efficiency appears to have been the key manufacturing competitive priority. From the mid-1960s on, quality also started to be considered a top priority. From the 1980s on, generally speaking, there has been a trend where, together with cost efficiency and quality, flexibility came to rank at the top of the competitive priorities of manufacturing companies. Some authors argue that in the 1990s time (delivery speed, quick product introductions and so on) is another top ranked criterion of many markets.

A more detailed analysis of manufacturing objectives, particularly flexibility, and their inter relationship can be found in the next section.

Manufacturing strategy and flexibility

In most of the cited manufacturing strategy work, flexibility seems to be regarded (at least implicitly) as having an important role in the organization's manufacturing strategy in at least two ways.

Firstly, as a response to an increasingly turbulent environment, flexibility can be seen as one of the most valuable features a company can possess regardless of the position which the company occupies in Hayes and Wheelwright's (1984) four-stage classification. If the company's manufacturing function still plays a purely reactive role, what matters is a good and quick reaction to changes in marketing needs, environmental and internal unexpected set-backs and so on. If, on the other hand the company's manufacturing function has a more proactive role, fast response to environmental changing conditions will lead to shorter response times, a feature which has been considered as one of the most important ones for the next decade's competitive battle.

Secondly, flexibility is very pervasive and can influence the performance of other competitive criteria of organizations. Slack (1989) argues that flexibility is a second order competitive criterion in the sense that a company does not win orders based on its flexibility as such, but based on other criteria (such as delivery time, reliability, cost or quality). The virtue of flexibility would be to support the achievement of the other, first order, competitive criteria. Slack's point is only partially right for, although flexibility can have an important role in supporting and influencing the effecting of the other competitive criteria, it can also be a first order competitive criterion. To illustrate this point, let us suppose a car manufacturer is developing a new model. If the approach adopted is that of simultaneous development (Womack et al., 1990), the company will probably have suppliers involved at the very early stages of the product development process, frequently even before the definitive specification of the parts is completely defined. If this is the case, changes in the preliminary specification tend to become a norm and it is likely that the hypothetical car manufacturer will prefer to choose a supplier on the grounds of its ability to respond effectively to such specification changes or in other words, based on the supplier's flexibility. This is an illustrative example of the potential of flexibility as a first order competitive criterion.

There follows a brief analysis of flexibility as a second order criterion or, in other words, the ways flexibility can influence the organization's other competitive criteria.

Generally, the performance criteria can be considered at both levels - the manufacturing system level and the resource level[6]. The relation between the level of performance of the particular resources and the level of performance of the system as a whole, regarding the same criteria, is not direct. That is because the effect of the interaction between the resources has also to be considered. The performance of a system in terms of a particular criterion is the aggregated effect of, on the one hand, the performance of its particular resources and, on the other hand, the interaction between the resources, with regard to the particular criterion.

Quality

There are two aspects or levels of quality: the level of quality which a machine can provide such as the tolerances it can work within and the scrap levels it normally produces; and the level of product quality provided by the whole system which is a function of the quality built throughout the whole process, including, for instance, design, material supply, transformation process and assembly.

Cost efficiency

A particular machine can provide a level of productivity and the whole system can provide a level of productivity which results in the cost of the products which the system manufactures. A system with very productive machines may have overall low productivity caused, for instance, by the operator's absenteeism, lack of skills or excessive levels of stocks.

Manufacturing speed

A particular machine can provide a level of speed, with fast processing times, and the whole system can achieve a speed which is reflected by how short are their overall production cycle times. A system with very fast machines and operators may have long lead times if its production planning system is inadequate, for instance, working with large lot sizes, causing queues to build up and causing throughput times to lengthen.

Manufacturing dependability

The particular resources have a level of dependability, e.g. given by a very low mean time between failures of a machine or the low absenteeism level of a worker, and the whole system has a level of dependability, reflected, for instance, by the delivery lead time dependability. A system with very dependable resources can have problems with dependability if it can not promise the delivery dates accurately or if it may not manage priorities properly in the plant, in case something goes wrong - which is inevitable. On the other hand a firm can achieve high levels of dependability if it develops system robustness (which includes contingency plans for unexpected set-backs, for instance) even though the individual resources are not extremely reliable.

Flexibility

The level of flexibility of the particular resources is rendered by the variety of tasks which they can perform and the ease with which they switch between tasks, and the level of the flexibility of the system as a whole is reflected by the effectiveness with which the system as a whole is able to change what is being done.

In order to change the levels of the manufacturing system's performance criteria an organization can adopt two main approaches:

i) Improving directly the level of the performance criteria of the main individual resources involved, for instance, by training people to 'do it right the first time' (aiming at quality improvements) or implementing a new machine which produces faster (aiming at improving delivery speed); and

ii) Improving indirectly the level of some performance criteria. Increased flexibility, particularly, can influence the level of the system in terms of dependability, cost, speed and quality (adapted from Slack, 1991).

Dependability can be enhanced by flexibility because a flexible manufacturing system is more apt to cope with unplanned or unexpected events affecting both process (such as machine breakdowns and labor absenteeism) and supply (such as faulty deliveries). As examples, flexible in-house processing capability is more likely to allow the production flow to be rerouted quickly and smoothly in case of equipment breakdowns. Labor flexibility can compensate for local shortages caused for instance by absenteeism because flexible workers can perform a variety of tasks and therefore can be transferred between work centers. A flexible manufacturing planning and control system can quickly regenerate schedules to better accommodate unexpected shortages in supply, 'diluting' the effects of the shortages.
Cost efficiency is improved by developing flexibility because a flexible system normally utilizes the resources more efficiently, with shorter process changeover (non-value adding) times. With shorter changeover times it is also

possible to work with smaller batches which reduce the levels and therefore costs of work in progress inventory. With smaller batches the production flow tends to be smoother, allowing for better utilization of equipment and people. A flexible system also tends to deal better with unexpected interruptions in the production flow, overcoming them with less organizational disruption and therefore being less costly. All these aspects can positively influence resource productivity and cost efficiency.

Speed, meaning fast delivery, fast development of new products, or fast customizing of products, can be improved by a flexible operation. Flexible changeovers give small batches, low work in progress, smooth production flow and therefore fast throughput; processes with a wide range of capabilities can accommodate new products without costly and time consuming new investments.

Quality can also be affected by flexibility. With a flexible system, the changeover operations are quicker and easier, making it also quicker and easier to bring production back to tolerances when a new production run starts. Therefore fewer sub standard products tend to be generated during the setup operations.

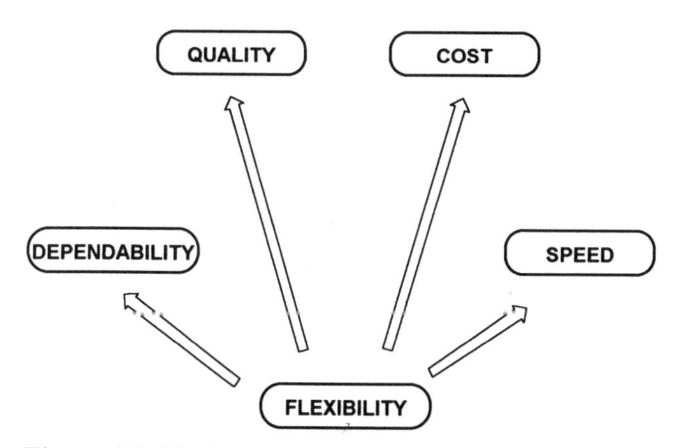

Figure 1.3 Flexibility's influence on other objectives

Flexibility thus deserves special attention firstly because its strategic importance as a first and as a second order competitive criterion is growing and also because the amount of research on the strategic aspects of flexibility has not matched its growing importance. The remaining discussion of this book is dedicated to flexibility. The main aspects presented in this Chapter with regard to flexibility will be further exploited in the next Chapters, particularly Chapters 2 and 3. The underlying assumption behind the rest of this work is that the main reason for an organization to seek manufacturing flexibility is to enhance its strategic competitiveness.

Notes

1 Economies of scope (Goldhar et. al., 1987) are said to occur when one production unit can produce a given level of outputs of a variety of products at an unitary cost which is lower than that obtained by a set of separated production units, each producing one product at the same level of output.

2 Hayes and Wheelwright (1984) propose the representation of the interaction between the stages of the product life cycle and the stages of the process life cycle using the product-process matrix. The rows in the matrix represent the major stages through which a production process tends to pass, going from what they call jumbled flow (job shop) to the continuous flow through disconnected line flow (batch) and connected line (assembly line). The columns represent product life cycle phases which go from the great variety associated with the product's initial introduction (low volume, low standardization) to the standardization, associated with commodity products (high volume, high standardization), passing through intermediate stages. The authors suggest that normally there is a 'natural match' between process and product stages in their life cycles, and that normally production systems would be located in the matrix according to diagonal matches, in which a certain kind of product structure (set of market demand characteristics) is paired with its 'natural' process structure (set of manufacturing characteristics).

3 Some empirical evidence can already be found, though, in the literature, e.g. in the work of the Boston Consulting Group, reported in Stalk and Hout (1990).

4 Hayes and Wheelwright (1984) propose a taxonomy for the manufacturing systems in which 4 stages are defined - internally neutral, externally neutral, internally supportive and externally supportive - according to the increasing proactive role played by the manufacturing function within the organization's corporate strategy.

5 Decision areas represent sets of decisions which relate to a specific aspect of the manufacturing function.

6 The system level is defined here as the level of the production units, or the set of manufacturing resources which interact, having general common objectives, as opposed as the resource level which is defined here as the level of the specific individual resources, e.g. an individual machine or a worker.

2 The flexibility of manufacturing systems: a review of the literature

Objective

The objective of Chapter 2 is to discuss the most relevant views of the authors in the literature about flexibility at the manufacturing system level. The manufacturing system is the level of the set of inter - related manufacturing resources which form the production units, as opposed to the level of the individual resources.

Initially, Chapter 2 extends the discussion initiated in Chapter 1 about the reasons for the growing view on the part of academics and practitioners that manufacturing flexibility is an important competitive criterion for the present and future markets. The recent publications which discuss manufacturing system flexibility are summarized and critically reviewed. The following aspects are addressed specifically:

i) the objectives of manufacturing flexibility: why organizations should or should not develop manufacturing system flexibility, according to the literature.

ii) the nature of manufacturing flexibility: several definitions and classifications of manufacturing system flexibility types and dimensions found in the literature.

iii) the methods of assessment: some methods for assessing flexibility proposed by various authors.

iv) the development of manufacturing flexibility: according to the literature the methods of developing manufacturing system flexibility.

Introduction

Several management research areas e.g. organizational behavior and industrial management have dealt with the concept of flexibility in recent years. Each of them approaches flexibility in a different way, depending on their scope. Each of them also uses different analytical tools, consistent with each approach. The organizational behavior researchers are mainly concerned with the flexibility of the human resources within the organization and in this sense they use tools such as behavioral theory, psychology and sociology of work, as can be found, for instance, in the work of the IMS (Institute of Manpower Studies) of Sussex University. In industrial management one can also find a vast literature on flexibility, mainly of the equipment involved in the production process, generally under the label FMS (flexible manufacturing systems), with a quite technical approach which focuses on issues like task sequencing or dispatching disciplines.

Although possibly useful for the solution of short-term area-specific problems, the kind of localized approach exemplified above seems to be inadequate for a more strategic and comprehensive understanding of manufacturing flexibility such as the one needed by the managers of the manufacturing processes. The approach adopted in this Chapter is comprehensive rather than partial, contemplating the set of inter related manufacturing resource types, rather than predominantly one or several.

The importance of manufacturing flexibility

Since the early 1980s, a new emphasis has been given to the importance of flexibility for the competitiveness of manufacturing systems. This new concern is based on a number of factors, identified by several authors and summarized below (Slack, 1987).

The environment in which the manufacturing companies have had to act has been extremely turbulent. Competitors have been more and more competent, the market has demanded an increasing variety of products with shorter life-cycles, and the suppliers have not always accomplished desirable levels of product quality and service level as well as prices, because they are struggling with their own difficulties in the same turbulent market. These conditions lead to a situation of limited stability and predictability and therefore demand an increasing capability to respond well to changing circumstances or, in other words, to develop flexibility.

The development of new process technologies. The development of new process technologies has been of such a proportion that the rate of technology development may have outstripped the ability to use it to its full advantage or even understand its potential (Voss, 1986). The result of this lack of balance would be the under utilization of the new technologies, which potentially offer technology flexibility to any organization that can manage to transform potential into actual flexibility. Great effort has been put by manufacturing organizations in trying to work out how to do it effectively.

Influenced perhaps by the development of the new process technologies - flexible automation, numerically controlled machines (either stand alone or integrated), automated material handling systems, pattern recognition systems and the flexible manufacturing systems, the concept of flexibility is frequently associated with technological resources, when discussed in the literature. However, although the importance of technology in the effort to achieve more flexible manufacturing systems may be recognized, it is important to bear in mind a broader perspective. Manufacturing systems consist of a set of resources, which, apart from technology, include people and infrastructural systems (e.g. the organizational systems and the supply systems).

The emphasis on technology flexibility is reflected in the work of some authors (e.g. Gupta and Goyal, 1989) according to whom, the introduction of flexibility into a manufacturing system requires high initial capital investment in flexible technology. A number of authors (Schomberger, 1986; Shingo, 1985; Blackburn and Millen, 1986) disagree with that view. They argue that a system can achieve flexibility by using simpler and cheaper machines as long as they are properly utilized, i.e. on the condition that their set-up times are sufficiently reduced.

Some authors have called attention to the risks involved in the myopic reduction of the broad concept of flexibility to the concept of flexible automation. Jaikumar (1986) reports the results of research realized with American companies which adopted FMS and concludes that most of them are actually inflexible because of lack of appropriate supportive managerial systems. On the one hand it would not be enough, according to this view, to have flexible automation to ensure the achievement of manufacturing system flexibility. On the other hand flexibility is not an exclusive characteristic of the automated FMS. Some authors, e.g. Poe (1987), argue that when real flexibility is needed the tasks have to be assumed by people. As an example, the author mentions the final assembly line in the automobile manufacturers where automation would be possible/viable for only 10 per cent of the tasks.

Bringing these two ideas together, it is important to be aware, when analyzing production systems, that although flexible automation may have an important role to play, it is not enough and in many cases not even necessary for the achievement of system flexibility. It is necessary that the whole set of structural (technological and human) and infrastructural (systems) resources are considered on a systematic rather than on a partial basis in order to develop a more comprehensive understanding of the flexibility of manufacturing systems.

The objectives of manufacturing flexibility

The authors identify a number of reasons for organizations to seek more flexible production systems. These factors can be separated into three principal groups:

i) factors related to the output side of the manufacturing system (consumers and competitors);

ii) factors related to the input side of the manufacturing system (suppliers); and

iii) factors related to the production process.

Objectives related to the manufacturing system's outputs

These are the most frequently mentioned factors in the literature. The markets are in a process of becoming increasingly fragmented, demanding more variety of products with shorter life cycles. This is happening in parallel with an increasing need for more efficiency of the processes and overall effectiveness of the manufacturing systems. The managers are facing the most turbulent market environment in many decades. The lack of predictability of the demand would be motivating companies to develop their ability to cope with this turbulence and to respond to it effectively (or, in other words, to improve their levels of flexibility).

The utilization of flexible resources is generally more costly than the utilization of dedicated ones. However, in a market which tends to demand diversification rather than mass production, one concept gains importance: that of the economies of scope (see Chapter 1), achieved through flexible systems.

The scarce resources which the organizations have available for developing flexibility make the resource allocation a problem of main importance. More than ever, it is important that the managers understand the concepts involved in such a decision process in order to be able to identify the most effective ways to invest in flexibility development.

Objectives related to the manufacturing system's inputs

Swamidass (1985) claims that 'the competitive value of manufacturing flexibility is its ability to neutralize the effects of demand uncertainties'. This claim does not seem to be confirmed by research work realized with a number of American, European and Japanese manufacturers (De Meyer, 1986): one of the findings of the researcher is that the American and European companies would not be adopting automated flexible manufacturing systems in order to be able to change their product designs quickly, as is broadly believed, but in order to accommodate the variability of their inputs. This tendency seems to be less evident with Japanese companies because the relationship they have developed with their suppliers would allow them to establish long term contracts and effective technological cooperation.

In environments where the supply market is less developed (e.g. some newly industrialized countries), the possible lack of reliability of the suppliers with regard to delivery time and raw material and component quality would motivate the development of more flexible production systems, which can cope with such imperfections.

In a similar way, in some markets, the supply of labor at the required levels of skills and quantities is uncertain. Companies operating in these environments normally try to develop more flexible manpower structures in order to be able to cope more efficiently with this kind of uncertainty through quick and easy re deployment and adjustments of manpower levels.

Another factor related to the supply of inputs is the development and availability of more flexible process technologies at lower costs. The development of these technologies is pushing the market to utilize them, either because these technologies represent the effective solution for an unresolved problem or because the non adoption of these technologies might result in loss of future competitive power. The cost of not acting is an important consideration when a company analyzes the adoption of new process technologies. Traditional analysis techniques such as discounted cash flow, return on investment and others have been considered insufficient to perform these analyses largely because they are unable to consider such strategic costs.

Objectives related to the production process

Flexibility would also be developed as an 'insurance' (Carter, 1986) against process uncertainty mainly in the short term. Equipment breakdowns and labor absenteeism are examples of uncertainties regarding the structural resources of the manufacturing process.

The nature of manufacturing flexibility

Throughout the literature, the authors seem to agree that flexibility is a concept which is not fully understood as yet. Several definitions are found, as well as several classifications of dimensions and types of flexibility.

Concept and definitions

One of the most often mentioned definitions of flexibility is still Mandelbaum's (1978): 'the ability to respond effectively to changing circumstances'. Despite its vagueness and its limited practical use, a brief analysis of the semantic elements of this definition can be used in an attempt to understand the difficulties involved in dealing with the concept of flexibility.

The first element of Mandelbaum's definition is 'the ability to...' which gives flexibility a character of a *potential*. This makes the task of measuring the flexibility of a given system both difficult and controversial. Analyses of the system's historical data are of limited use and do not give anything but an idea of its ability to respond to changing circumstances. The system's past performance may be close to the limit of its ability but it may also not be. The system could for instance, be able to cope with more and/or more dramatic changes than it has done in the past. These aspects make objective measuring of flexibility a very difficult and controversial task.

The second element is '...to respond...'. Response generally means reaction or adaptation to some sort of change, given that the change has already occurred.

The third element of Mandelbaum's definition is '... effectively...' , which suggests a link between the concept of flexibility and the concept of overall system's performance. In other words, this means keeping (despite the change) or improving (responding to the change) its overall performance.

The last element of Mandelbaum's definition is '...changing circumstances...'. Behind this element there are two concepts: uncertainty and variability. On the one hand if the circumstances are changing but the steps of the change are perfectly known (therefore without uncertainty) one could have more time to plan the means to cope with the change. Still, the more the circumstances change the more flexibility is demanded from the system in order to cope with them. On the other hand if the circumstances are recognized as changeable or changing and on top of that one does not have perfect information about the result or the steps of the change (therefore with uncertainty), then it is necessary that the organization develops extra abilities such as to be able to identify several factors: what is the extension of the possible uncertain changes, what part of this range is involved and how effectively does the organization intend to be able to cope in order that the appropriate levels of flexibility can be planned and achieved.

The possible uncertainty behind the changing circumstances adds difficulties to the treatment of flexibility. When one discusses uncertainty regarding the market or labor behavior, for instance, the statistical and mathematical analytical tools frequently used to model and analyze some simpler uncertain events become of limited use. Other methods, some of them borrowed from the social sciences, may have to be used.

There are a few other definitions of flexibility in the literature which are conceptually different from Mandelbaum's. Ferdows and Skinner (1986) propose a different approach: according to the authors' view, flexibility should be seen as a relative variable. A flexible system would be more able to react quickly and at lower costs than the competitors to the market changes. This view does not seem to be appropriate because it means to mix a variable itself and its value. It may be useful to understand and analyze a system flexibility independently of its relative position when compared to the companie's competitors. Although it may be very important competitively, a company competes based not only on its flexibility but on a set of criteria which may include cost, quality, and delivery performance among others. A company may, for instance, choose to be less flexible than the competitors, provided that its level of flexibility reaches the minimum level required by the market, and prefer to compete in other terms such as design leadership. In this sense, it seems to be more appropriate to consider flexibility as an absolute concept as opposed to relative. Slack (1990) also seems to believe that flexibility is not a relative concept since he argues that 'the flexibility of the operation as a whole is determined exclusively by the flexibility of its constituent resources and systems.'

Dimensions and types

Some authors argue that flexibility is a multi- dimensional variable. Slack (1983) defines two basic dimensions: range and response. *Range* flexibility would be the ability of the system to adopt different states. One production system will be more flexible than another in a particular aspect if it can handle a wider range of states, for instance, to manufacture a greater variety of products or to produce at different aggregate levels of output.

However, adds Slack, the range of states a manufacturing system can adopt does not totally describe its flexibility. The ease with which it moves from one state to the other in terms of costs, time and organizational disruption is also important. A production system which moves quickly, smoothly and cheaply from one state to another should be considered more flexible than another system which can only cope with the same change at greater cost and/or organizational disruption. The way the system moves from one state to another would define Slack's other flexibility dimension, *response* flexibility.

Although range and response are clearly two different dimensions of flexibility, it is important to notice that they are not independent. Manufacturing systems tend to be more responsive to small changes and less responsive to big changes (Slack, 1989).

Time is another dimension which to some authors is important for the understanding of flexibility. Carter (1986) believes that different kinds of flexibility have an impact on the production system in different time frames: very short term, short term, medium term and long term; as a consequence, different kinds of flexibility should be sought in order to achieve the different time frame objectives.

Stecke and Raman (1986) also consider time in their analysis regarding the relationship between flexibility and productivity and propose that, in the short term, production flexibility enables the system to maintain its production in face of unforeseen events, such as machine breakdowns. With regard to the long term, Stecke and Raman propose that production flexibility would be related to the inter-dependence between the process and product life cycles. Flexible systems in the long term would tend to cause a relaxation in the one-to-one relationship which the conventional production systems would represent. This relationship is discussed in detail in Hayes and Wheelwright (1984), Chapter 4.

Another dimension identified by Gerwin (1986) as a basic issue in defining manufacturing flexibility is the level at which it is to be considered:

i) the individual machine or manufacturing system;

ii) the manufacturing function such as forming, cutting or assembling;

iii) the manufacturing process for a single product or group of related ones;

iv) the factory or,

iv) the company's entire factory system.

At each level, says Gerwin, the domain of the flexibility concept may be different and alternative means of achieving flexibility would therefore be available.

A company which intends to be flexible in the introduction of new products in the market place (at the highest level, that of the company's entire factory system) should take actions different from those of a company which intends to make a machine more flexible by developing jigs and fixtures in order to shorten its set-up time (the lowest level, of the individual machine). In the former for instance, it is essential that the flexibility of the product design team is developed. In the latter the flexibility of this team is possibly less important.

Gupta and Buzacott (1986) define three dimensions of manufacturing flexibility: sensitivity, stability and effort. With respect to each change, sensitivity relates to the magnitude of the change tolerated before there is a corrective response. Stability relates to the size of each disturbance or change for which the system can meet expected performance targets. Whereas sensitivity and stability determine whether a system responds to a change or not, effort relates to how well a system responds to a change. Effort depends on such factors as the time to respond to a change and cost of response.

Dooner and De Silva (1990) propose dimensions which are similar to Slack's. According to these authors, flexibility would have three dimensions: range, switchability, and modifiability. Range, similarly to Slack's range, relates to a set of states a machine or a set of machines can adopt to do useful work. Within a given set, transitions can be made between states. The general ease with which this takes place is called switchability. Modifiability would relate to taking up a new set of states, which may or may not include those individual states belonging to the set of states prior to the modification.

Mandelbaum (1978) defines two basic dimensions of manufacturing flexibility: action flexibility and state flexibility. Action flexibility would be the capacity for taking new action to meet new circumstances, that is, leaving options open so that it is possible to respond to change by taking appropriate action. State flexibility would be the capacity to continue functioning effectively despite the change, i.e. the system robustness or tolerance to change.

Figure 2.1 summarizes the different dimensions of manufacturing flexibility according to selected authors

Mandelbaum (1978)	Slack (1983)	Gupta & Buzacott (1986)	Stecke & Raman (1986)	Carter (1986)	Gerwin (1986)	Dooner & De Silva (1990)
action	range	sensitivity	time	time	organization-al level	range
state	response	stability				switch-ability
		effort				modifi-ability

Figure 2.1 Dimensions of manufacturing flexibility according to some selected authors

There are similarities and differences between flexibility dimensions found in the literature. Gupta and Buzacott's effort seems to be quite similar to Slack's response. Slack's concept of response is divided by Dooner and De Silva into switchability and modifiability. Slack's range is similar to that of Dooner and De Silva. In addition, Slack's range is similar to Gupta and Buzacott's stability but their sensitivity, as a flexibility dimension, is somewhat arguable mainly if we consider flexibility as the ability to respond to change. Sensing the changes depends on the ability of the system to monitor the changes rather than on the ability to respond to the changes (which is proper flexibility).

Stecke and Raman's and Carter's time dimension and Slack's, Dooner and De Silva's and Gupta and Buzacott's dimensions do not seem to be independent. Short term considerations seem to be associated with response, effort and switchability whereas long term considerations seem to be more associated with range, sensitivity and stability and modifiability because they are more related to structural changes such as the resources and their capabilities.

Gerwin's organizational level considerations seem to be very important for defining the boundaries of the system which is being analyzed. One of the difficulties one finds in analyzing the literature on flexibility is exactly the fact that the authors not always define the scope or the organizational level which they are considering.

With regard to the flexibility types, the classifications found in the literature vary according to the approach which each particular author adopts. All of them seem to agree that classifying flexibility into different types is important. This suggests that different kinds of flexibility would be obtained by different means (possibly developing the different resources in different ways) or would be appropriate for dealing with different conditions or types of change.

Buzacott (1982) defines two types of manufacturing system flexibility, based on the change the system has to cope with: job and machine flexibility. Job flexibility would be the ability of the system to cope with changes in the jobs to be processed by the system. Machine flexibility would be the ability of the system to cope with changes and disturbances in the machines and at the work stations.

Zelenovic (1982) proposes two types of flexibility: design adequacy and adaptation flexibility. Design adequacy is the probability that the given structure (machines, handling equipment, measuring equipment, storage and control devices and plant layout) of a production system will adapt itself to the changing environmental conditions and to the process requirements within the limits of the given design parameters (of the given structure). Adaptation flexibility is the ability of the system to transform/adapt from one to another job task at minimum value of time. Zelenovic's adaptation seems to be similar to Slack's response dimension.

Slack (1988) suggests four types of manufacturing flexibilities which would be achieved through the development of flexible resources:

i) product flexibility: the ability to develop or modify products and process to the point where regular production can start. If range is considered, this is similar to Zelenovic's design adequacy;

ii) mix flexibility: the ability to produce a mix, or change the mix of products within a given time period;

iii) volume flexibility: the ability to change the absolute level of aggregate output which the company can achieve for a given product mix; and

iv) delivery flexibility: the ability to change delivery dates effectively.

Dooner and De Silva (1990) also consider four types of manufacturing flexibility:

i) machine flexibility: the ability of a machine to accommodate different tasks;

ii) mix flexibility: the ability of a system to accommodate different types of part design which can be manufactured simultaneously;

iii) part flexibility: the ability of a system to accommodate new or modified part designs; and

iv) volume flexibility: the ability of a system to accommodate variations in the production rate.

Gerwin (1986) defines seven types of flexibility of production systems as part of his attempt to establish guidelines to the relationship between different types of uncertainty to which the organization is subjected and the types of flexibility which the company should use in order to cope with them. Gerwin's flexibility types are:

i) mix: the ability of a manufacturing process to produce a number of different products at a certain point in time.

ii) changeover: the ability of a process to deal with additions to and subtractions from the mix over time.

iii) modification: the ability of a process to make functional changes in the product.

iv) re-routing: the degree to which the operating sequence of the flow of the parts can be changed.

v) volume: the ease with which changes in the aggregate amount of production of a manufacturing process can be achieved.

vi) material: the ability to handle uncontrollable variations in the composition and dimensions of the parts being processed.

vii) sequencing: the ability to rearrange the order in which different kinds of parts are fed into the manufacturing process.

Although similar to Gerwin's in many aspects, Browne et al. (1984) proposed another way to classify the flexibility of production systems in types. Stecke and Raman (1986) also use this classification in order to analyze the relationship between the flexibility of production systems and their productivity. This classification is as follows:

i) machine: the ease with which the operations of a given set of part types can be performed at a given machine.

ii) process: the ability of the manufacturing system as a whole to manufacture a given set of part types in several ways.

iii) routing: the ability of a system to maintain its efficiency in the face of breakdowns.

iv) operations: the ease with which the sequence of operations for each of the given part types can be inter changed.

v) volume: the system's capability to be operated profitably at different volumes of the existing part types.

vi) product: the ability of the given manufacturing system to changeover efficiently from a particular set of part types to a different set.

vii) expansion: the manufacturing system's capability to be built and expanded modularly.

viii) production: the cumulative result of the seven previous flexibilities.

Most of these classifications do not seem to give the same emphasis or importance to all the resources involved in the production process. The technological resources seem to deserve a much greater attention in the analysis than the human and infrastructural (information and managerial systems) resources.

There is no standardization in the terminology about flexibility matters in the literature. Mix flexibility, for instance, means different things to Slack (1989) and Gerwin (1986). That makes it difficult to do comparative work between the authors' classifications. Another factor which was highlighted by Gerwin (1986) and is in general not explicitly stated by the other authors who propose the different classifications is the organizational level they are considering. Some classifications in the literature mix flexibility types of two or more levels. Browne et al.'s (1984), for instance, includes machine flexibility (individual resource level), process flexibility (manufacturing system level) and expansion flexibility (company level).

From the classifications listed above, it can be seen that there are some similarities between the authors' flexibility types. Figure 2.2 shows five authors' flexibility types. Types which bear some similarity are shown in the same row. Zelenovic's (1982) design adequacy and adaptation seem to be flexibility dimensions rather than types as defined here, therefore they are not shown in Figure 2.2.

Figure 2.2 shows the lack of consistency between the selected authors in terms of terminology and approach. Slack's seems to be the classification which is more directly associated with the manufacturing system's strategic objectives, since the four types which he suggests are consistently at the manufacturing system level and refer directly to the system demand. The only flaw, possibly, is that it focuses exclusively on the system's output side, neglecting the flexibility component which can be used to overcome process and input set-backs. This component mentioned by a number of authors (e.g. Stecke and Raman, 1986 and Carter, 1986), may have strategic implications. This point will be further analyzed in Chapter 4.

Buzacott (1982)	Browne et al. (1984)	Gerwin (1986)	Slack (1988)	Dooner & De Silva (1990)
	product	mix changeover	mix	mix
		modification	product	part
	volume	volume	volume	volume
		sequencing	delivery	
machine	routing	re-routing		
	machine			machine
		material		
	process			
	operations			
	expansion			
	production			
job				

Figure 2.2 Types of manufacturing flexibility according to some selected authors

The measures of manufacturing flexibility

One of the difficulties found by the authors who study the flexibility of manufacturing systems is how to measure it. Two main streams can be identified among the authors. There are the ones who seek to define objective measures and the ones who prefer to assess the flexibility based on the perception of the people involved in the process.

Some authors (e.g. Slack, 1988; Swamidass, 1985) seem to prefer perceptive measures while others (Stecke and Raman, 1986; Zelenovic, 1982; Gerwin, 1986) seem to prefer the objective methods. All of them agree however that it is important to have a procedure in order to assess the flexibility of a system or, at least, the flexibility needs of a system and that this assessment should be done periodically in order to face an increasingly dynamic environment.

Objective measures

One of the main difficulties in the development of objective measures of flexibility is its characteristic of a potential (Tidd, 1991). Flexibility would be an ability, a potential to realize things rather than something measurable with hindsight, such as performance.

Gerwin (1986) suggests that, in order to measure *modification flexibility*, one should reckon the number of design changes made during a period in one component. This does not seem to be appropriate because the number of design changes which have been made may mean that the market demanded only such number, rather than that the number of the design changes was limited by the ability of the system to realize them. Thus one could say that the ability of the system to realize design changes is at least equal to the number of changes realized in any one period but no one can assure that the system could not have realized more design changes than the number reckoned at the end of the period.

Another operational problem with regard to objective measures of flexibility is that when one talks, for instance, about the number of design changes made one is leveling the treatment of changes which might be completely different from each other in terms of magnitude or complexity.

Two different hypothetical production systems could realize, during the same period of time, the same number of changes (which could even be similar in magnitude and complexity). However, if one of them performs the changes more easily (with less cost, time or disruption) than the other, it seems to be reasonable to consider it more flexible than the other. Based on Gerwin's measure both systems would show the same flexibility level. Schmigalla *apud* Zelenovic (1982) proposes an index to measure flexibility which has the same operational problems as Gerwin's and is restricted to *machine flexibility*. The index uses, for example, a variable 'K_{ei}' which represents 'the effective capacity of machine 'i''. However, it would be difficult to decide what effective capacity to use in the index, for flexible machines are able to manufacture several different products and in many cases, the capacity of a machine varies according to the product it is currently making.

Kumar (1987) proposes a method to assess flexibility using the concept of entropy. The method is based on the alternative choices which a system has available and on the *reliability* of each choice. Reliability in this case is defined as a measure of the relative preference which the different choices would deserve. The bigger the number of possible choices and the more similar the preferences between them are, the higher the flexibility indicator will show. Again, the problem with this kind of numerical indicator is that it requires that the preferences of the possible choices are quantifiable.

Kumar's method is interesting and possibly useful in order to compare the flexibility of sets of machines, for example. However, with regard to assessing manufacturing system flexibility as a whole, where there is a concurrence of several different types of resources, it is difficult to rank preferences or even to identify and quantify possible choices. Besides, the measures based on entropy cannot capture the responsiveness of the systems: two systems can have the same number of options available with the same set of relative preferences. They would be given the same flexibility indicator value. Yet, one system could, with the same set of choices, be more responsive than the other, being therefore more flexible.

The authors who try to find objective methods to measure flexibility contribute to enlightening of a not sufficiently explored subject. However, when one seeks to develop models in order to support the decision making regarding production systems, oversimplified indexes and measures may, in a dangerous manner, fail to hold to the modeled reality. The most serious problem about these oversimplified measures is that, once the index is defined by one author, people who will apply it sometimes do not seem to pay enough attention to the hypothesis assumed when the index was developed. Sometimes, in production management environments, a decision seems to gain an overrated legitimacy when it is, and just because it is, based on a mathematical, quantifiable expression. Its use sometimes becomes indiscriminate and, in this case, the resulting decisions may be wrong.

Perceptual measures

For the reasons discussed in the previous section, in terms of measuring complex variables such as the flexibility at the manufacturing system level, the methods which use the perception of experienced people involved in the process, when well conducted, can have advantages over the quantified, hard data-based ones.

Slack (1988), based on Fine and Hax (1985), proposes a method to assess the flexibility of production processes which is based on the perception of the managers involved. The method uses scales which consider the relative position of the assessed system among competitors. For each type of flexibility (product, mix, volume and delivery) and for each of the dimensions (range and response) the production system is classified into nine Likert type categories ranging from 'very much better than the nearest rival in industry' to 'the lowest capability in industry'.

The managers' scaling is then compared with another scaling where the managers point out the capability levels considered important to the competitiveness in the market. The gaps between the actual capabilities and the important capabilities guide the decision making of the managers. The answerers should be experienced people capable of developing a global rather than a partial view of the organization. An alternative to comparing the company's flexibility performance to the competitor's performance is to compare the company's flexibility performance with the customer's (which can be internal or external) expected levels of flexibility.

Complementary techniques can be used in order to arrive at a consensus assessment rather than at an average one. The answers should, as much as possible, be based on data (objective as well as subjective) which then should be provided. Slack (1988) suggests the use of accessory tools such as the range/response curve, in order to help the assessment of product flexibility, for example. This way, no information would be overlooked and, on the other hand, the analysis would not be restricted to the numerical data available.

The development of manufacturing flexibility

One of the most discussed aspects of the literature about flexibility is exactly how to develop it. The manufacturing system flexibility is achieved through the development of some specific characteristics of the manufacturing resources - people, technology and systems. The authors seem to agree that developing flexibility is in general desirable. Adler (1987) though warns: flexibility is a characteristic which people have tried to develop in production processes in order to make them able to cope with unstable and/or unpredictable situations. However, if the next decade brings, as seems plausible, new and more stable configurations of supply and demand some flexibility efforts may prove to have been myopic over-reactions. In other words, if the reasons for being flexible are eliminated, why be flexible?

Although Adler's point is worth considering, to a certain extent, the present preoccupation of managers and academic researchers about flexibility can be justified: on the one hand different countries and even different production systems are in different stages of development, in terms of the relations between the organizations and the market. While, for instance, in Japan the efforts to reduce the uncertainty of the supplier-customer relationship (e.g. through long term contracts and cooperation) seem to be at an advanced stage, the same does not seem to occur with regard to most of the western countries and particularly with the developing countries. In such matters, although the tendency is apparently that companies are trying to reduce their uncertainties, a cooperative, stable and predictable environment cannot be achieved overnight. In the meantime, the companies must develop ways to cope with their uncertainties, and developing flexibility is one of such ways (Swamidass and Newell, 1987).

On the other hand Adler's caveat may be more suitable for the uncertainties and instabilities related to the supply market. Concerning the consumer market, most authors agree there is a tendency that, in the future, it will demand an increasing variability of products with shorter life cycles (Slack,

1991; Gupta and Goyal, 1989; Gerwin, 1986). This, together with an increasingly competitive environment, may in itself justify the present efforts for the development of flexibility. See Corrêa (1992) for a detailed treatment of flexibility at the resource level.

Summary and conclusions

The literature would indicate that flexibility is not a fully understood concept as yet, surely deserving further research. The amount of recent articles on the concept of flexibility and its classification into types and dimensions attests this fact. However, flexibility has been considered by most of the authors in the literature as one of the most important characteristics of the manufacturing organizations in today's and tomorrow's market. The literature identifies two main reasons for this newly found importance: firstly, the environment in which the manufacturing companies have had to act has been extremely turbulent, demanding an increasing ability of the systems to respond to changing circumstances. Secondly, the development of new process technologies has increased the availability of machinery which embodies features such as flexible automation. However, it is important to be aware that although flexible automation may have an important role to play, it is not sufficient to ensure that the manufacturing systems will achieve flexibility, and in many cases not even necessary for the achievement of flexibility.

The literature can be divided under four main headings, according to the main focus with regard to manufacturing flexibility: focus on the objectives, on the nature, on the assessment and on the development of flexibility.

In terms of objectives, the authors consider that flexibility is developed mainly in order to cope effectively with uncertain and variable changes, whether they are environmental or internal, or related either to the inputs, to the outputs or to the manufacturing process. Yet, the literature lacks proper contingency models which relate what types of flexibility should be used in order to cope with what types of uncertainty and variability. Chapter 4 deepens this discussion and proposes an original analytical model to understand and analyze manufacturing flexibility.

With regard to the nature of manufacturing flexibility, several authors propose ways of classifying flexibility into types and dimensions. There is still a lack of commonly accepted terminology in the literature regarding the different types and dimensions of flexibility. One of the basic issues in terms of flexibility dimensions seems to be the level of analysis at which flexibility is to be considered. See Chapters 4 and 5 for further discussions on the nature of manufacturing flexibility.

In terms of measures for manufacturing flexibility, two main approaches can be identified in the literature. The first approach seeks to find objective, numerical measures whereas the second prefers to assess flexibility based on the perception of the people involved in the process. In terms of assessing the flexibility of the manufacturing system as a whole, the perceptual measures may be more appropriate because of the risks of oversimplification with the hard data-based, quantified measures.

3 Uncertainty and variability in manufacturing systems: a review of the literature

Objective and summary

The objective of Chapter 3 is to discuss the views found in the literature with regard to the uncertainty and variability that affect manufacturing systems.

Initially, the concept of uncertainty is presented and an important issue is addressed: is uncertainty an objective or a perceived category? A discussion follows on the methods of assessing uncertainty as they are found in the literature, with regard to both approaches: perceptual and objective.

The second part of Chapter 3 considers variability which, as the literature suggests, involves another reason for manufacturing systems to have flexibility: the variability of the outputs. There is an additional discussion, based on the literature, on the benefits and also the costs incurred by a manufacturing system which intends to provide high levels of variability to its customers. The reasons for product proliferation are explored. The implications of the current literature for this research are also discussed.

Concepts

Environmental uncertainty is one of the main reasons for a firm to seek flexibility (Swamidass and Newell, 1987; Gerwin ,1986; Slack, 1989). Uncertainty is a term which is used daily in a variety of ways. This everyday acquaintance with uncertainty can be seductive in that it is all too easy to assume that one knows what he is talking about. However, there has not been a consensus among the authors who have dealt with this complex and important subject on the concept of uncertainty. Gifford et al. (1979), reviewing the literature on uncertainty, identify considerable diversity in the terminology used by various authors, but they find two general notions which characterize the various approaches - information load and pattern/randomness.

The first, information load, is related to the complexity of the decision situation. The second category distinguishes between patterns and randomness of events. The classical definition of risk as the ability to assign probabilities to outcomes and of uncertainty as the inability to assign these probabilities (Luce and Raiffa, 1957), is based on differing perceptions of the existence of orderly relationships or patterns. The two concepts are not necessarily independent. Considered together, these two concepts imply that uncertainty will be low if data are available at the time needed and if the decision maker discerns a pattern of regularity among the cues that makes the data become useful information.

Lawrence and Lorsch (1969) suggest that environmental uncertainty is composed of three elements: lack of clarity of information, general uncertainty of causal relationships between decisions and the corresponding results, and time span of feedback about the results of the decision. All the uncertainty elements seem to refer to information which could help predict future events and/or trends. It is important to notice that even high coefficients of variation of the future events and trends do not necessarily indicate that the firm cannot predict them. It is the deviation from the expected which is important in regard to uncertainty and not the size of the range of events or the trend itself (Downey et al., 1975).[1]

Nevertheless, the existence or non existence of information itself, resulting from stimuli from the environment, is not the only factor that influences the level of uncertainty under which an organization operates. The set of stimuli lacks meaning or information value until it is perceived by an individual. Perception refers to the process by which individuals organize and evaluate stimuli (Secord and Backman, 1964). According to Huff (1978) the more one considers the notion of relative definitions of reality, individual values and experimental learning, the more important it is to look at the respondent to uncertain environments.

This is a point of great importance because it reveals a controversial subject in the literature: is uncertainty perceived or objective?

Uncertainty: perceptual vs. objective

Some authors propose objective measures for uncertainty based on physical attributes of the environment such as technological factors, number of product changes or research and development expenditures; these would be the indexes of uncertainty. On the other hand, depending on the previous level of knowledge of the individual and his cognitive process, the same set of stimuli from the environment can foster different levels of perceived uncertainty in different individuals. This is why certain entrepreneurs can, for instance, predict market behavior more accurately than others (many times, based on similar data), and therefore work under less uncertainty. What is certain to one person is uncertain to another (Huff, 1978). Environments, therefore, are neither certain nor uncertain but are simply perceived differently by different organizations (Perrow, 1967).

The restriction of uncertainty to a perceptual concept contains the inherent problem that some variations in perceived uncertainty are related to characteristics of the individual. It does not, however, preclude the expectation that uncertainty is also related to certain environmental attributes. In this sense, Downey et al. (1975) argue that specific attributes of physical environments tend to elicit similar perceptions of uncertainty by individuals. However, according to the authors, these similar perceptions of uncertainty by individuals stem from similarities in individual perceptual processes rather than from the existence of uncertainty in the physical environment. Downey and Slocum (1975) propose that perceived uncertainty can be expected to vary with:

i) Perceived characteristics of the environment;

ii) Individual differences in cognitive processes;

iii) Individual behavioral response repertoires; and

iv) Social expectations for the perception of uncertainty.

Perceived characteristics of the environment Duncan (1972) tried to establish relationships between the managers' perceptions of uncertainty and some characteristics of the environment. He characterized environments along two dimensions - *dynamism* and *complexity*. In Duncan's definition, a dynamic environment is one in which the relevant factors for decision making are in a constant state of change. A complex environment is one in which the number of interactive relationships relevant for decision making require a high degree of abstraction in order for manageable mappings to be produced by the individual.

Assuming that the perception of complexity/dynamism in the environment can be expected to be positively related to the perception of uncertainty, Duncan (1972) developed an instrument to measure environmental uncertainty and concluded that individuals in decision units with dynamic-complex environments experience the greatest amount of uncertainty in decision making and that the static/dynamic dimension of the environment is a more important contributor to uncertainty than the simple/complex dimension.

Individual cognitive process. Different individuals process stimuli from the environment in different ways. Duncan suggests, for example, that individuals with a tolerance for ambiguity may perceive situations as less uncertain than do individuals with lower tolerances.

Behavioral response repertoire. The individual's capacity to display appropriate behavioral responses to given environmental characteristics. Such capacities would not include an individual's innate qualities but rather capacities stemming from the individual's acquired experience.

Social expectations. A socially learned component in the individual's tendency to perceive uncertainty. For example, the degree of discretion defined for a position or person might be considered as an indicator of the organization's expectation of that position or person regarding uncertainty. Stated differently, if an organization expects a position incumbent to display little discretion, it can be assumed that the organization also expects the occupant to perceive little uncertainty.

Measurement

Many authors have tried to develop ways of measuring perceived uncertainty. Lawrence and Lorsch (1969) pioneered the effort developing a nine-item questionnaire designed to measure uncertainty in the three sub-environments of marketing, manufacturing and research within organizations: the respondent is asked three questions about each of the sub-environments. The response to each question is evaluated using a Likert-type scale. The questions and the response categories determine the extent to which each sub-environment is perceived as uncertain according to the following characteristics:

i) Lack of clarity of information;
ii) General uncertainty of causal relations, and
iii) Long time span for feedback of results.

Tosi et al. (1973) tested and analyzed Lawrence and Lorsch's instrument. In general, the result of their analysis did not reflect favorably on Lawrence and Lorsch's instrument. Replying to the criticism in a later article, Lawrence and Lorsch, although regretting not being enthusiastic about the contributions of Tosi et al.'s article, admit that there remains a need for methodological improvements in characterizing and measuring the environmental uncertainty of organizations.

Duncan (1972) attempted to develop another instrument for measuring uncertainty on the basis of three characteristics: i) the lack of information regarding environmental factors associated with decision making situations, ii) the lack of knowledge about the organizational consequences of a decision if the decision is incorrect, iii) the ability or inability to assign probabilities as to the effect of environmental factors on the success or failure of the organization performing its function. The first and second dimensions are measured by six and five Likert-type questionnaire items respectively. The last part is measured by a single two part questionnaire item. Each of the parts is scored in a specific way and the scores on the three characteristics are summed up to derive a total uncertainty score.

Downey et al. (1975) studied 51 managers of a U.S. conglomerate in order to examine the conceptual and methodological adequacy of Duncan's and Lawrence and Lorsch's uncertainty scales. Both methodologies were applied, compared and confronted with some criteria uncertainty measures. The principal findings of the authors were:

First, there appears to be a lack of communality between the two uncertainty scales which were presumably designed to measure a similar, if not the same, concept. Second, the total uncertainty scales did not correlate highly with any one of the four criteria uncertainty measures except for one. Several reasons are pointed out for the lack of correlation, the principal of them being:

i) the uncertainty sub-scales may not meaningfully lead to a total uncertainty scale because of, for example, inappropriate conceptualization of the uncertainty multidimensionality;

ii) the criterion uncertainty measures may not be appropriate indicators of perceived environmental attributes; and

iii) the relationship between specific behavioral environmental elements and specific characteristics which are usually considered as uncertainty sub-dimensions may be better understood than relationships between the global concepts of environment and uncertainty.

The authors add that if the latter reason is accepted, the analyzed instrument is of little explanatory or predictive value and should be restricted to more pedagogical purposes. Downey et al. (1975) suggest that considerable care should be exercised in selecting existing instruments for uncertainty measurement. The researcher should be sure that the uncertainty concept implicit in the selected instrument is consistent with the uncertainty conceptualization, either implicit or explicit, which is guiding the research.

Setbacks of perceptual-based uncertainty measurements

When considering perceptive measures - as opposed to objective ones - one should be aware of some of the mechanisms people customarily use to make judgment under uncertainty. Tversky and Kahnemen (1989) analyzed the way people make judgments under uncertainty and showed that, when trying to assess the probability of an uncertain event or the value of an uncertain quantity, people rely on a limited number of heuristic principles which reduce the complex task of assessing probabilities and predicting values to simpler judgmental operations. These heuristics, although generally quite useful, can sometimes lead to severe and systematic errors.

Conclusions on the analytical treatment of uncertainty

Considerable diversity can be found in the literature, in the terminology used with regard to uncertainty. One of the most controversial points in the literature is the discussion on whether uncertainty is a perceived or objective category. Some authors propose objective measures for uncertainty, which are based on attributes of the environment as indexes of uncertainty. However, it is argued in the literature that, depending on the previous level of knowledge of the individual and on his cognitive process, the same set of stimuli from the environment can foster different levels of perceived uncertainty in different individuals. According to this view, the set of stimuli lacks meaning or information value until it is perceived by an individual.

It seems to be reasonable to assume that uncertainty should be considered a perceptive rather than an objective category for the purposes of this research work for the following reasons:

First, a lack of correspondence between publicly available objective indicators of environmental change and managers' reports of perceived uncertainty has been noticed (Gifford et al., 1979).

Second, the environmental changes represent stimuli which are the cues or messages that are potentially available to the decision makers. In laboratory situations the term objective uncertainty is useful in describing stimulus conditions that are under control of the experimenter, such as the stated probability of winning a gamble. Outside of a carefully controlled laboratory situation, the nature of the cues utilized by decision makers cannot be easily specified, and there appears to be little value in the use of the term objective uncertainty (Gifford et al., 1979).

Environmental uncertainty is seen by a number of authors in the literature (Swamidass and Newell, 1987; Gerwin, 1986; Slack, 1989; among others) as one of the main reasons for a firm to seek flexibility. However, few of them have actually presented a detailed discussion on the relationship between, for instance, different types of uncertainty and different types of flexibility. Some of these authors seem to avoid discussing uncertainty in more detail, perhaps because of the difficulties involved in dealing with such a complex category. In fact, the organizational behavior theorists, who seem to be responsible for most of the research on the environmental uncertainty categorization and measurement, have produced results of little use for the manufacturing management field.

As a result, surprisingly, in the increasingly turbulent environment of today's competitive market, the literature on manufacturing management contains little research work which considers environmental uncertainty explicitly. Certainly, more research work is needed in the area.

In terms of the objectives of this research and for the sake of simplicity, the most appropriate approach seems to be that of Gifford et al. (1979). According to this author, considered globally, uncertainty will be low if data are available at the time needed and if the decision maker discerns a pattern of regularity among the cues that make the data useful for the prediction of future events or trends. The idea of uncertainty, according to this view, is broadly associated with that of *predictability*. Predictability seems to be a concept which is less controversial than uncertainty and also closer to the jargon normally used in industrial environments and therefore probably more easily understood by interviewees in such environments during the field work stage.

Variability in manufacturing systems

When analyzing the variability of outputs of manufacturing systems, at least two different dimensions can be envisaged: one is the actual variety of outputs, which refers to the range of products the system produces, and the other one is the variation of the system's outputs during the time period, not only in terms of how the range of products varies (e.g. product introductions or changes and breadth of product range) but also in terms of how the volume (aggregate and per product), the mix and the timing of the demanded output vary during the time period.

The costs of variability

The costs of an organization are in general very sensitive to the amount of variety. According to Miller (1988), the cost of variety is greater than the accounting system reports it to be. For the author, variety creates a cost burden throughout the company. Among other factors, the purchasing department, the materials management, vendor control and assessment, the incoming inspection and the stock of incoming materials are more costly as variety increases; the production lines are neither able to get up to speed to establish learning curve rhythms nor to obtain dedicated machinery to optimize the runs, as the setups are more frequent; the quality assurance function must develop separate standards and procedures for the different processes and various products; the investment required for both process and finished inventories is increased by variety.

Although it is arguable that all the mentioned costs would increase with product variety, considering primarily the new flexible technology, it is clear that some of them would. Let us now draw from Stalk and Hout (1990) and imagine the simplest of organizations:

> [...] an organization which manufactures only one product for only one customer [...] One product made day in and day out. Since there would be no changeover, production time lost to set-up would be negligible. Since there would be only one product, each step of the process would have matched capacities and be operated in unison. Quality costs would probably be low, since the process would remain unchanged and the quality problems would have been worked out. Inventories would probably be very low since purchased items could be brought in regularly, work-in-process kept to a minimum, and finished goods shipped immediately to the customer. Management costs would be low because everything would be almost perfectly predictable.
>
> Unfortunately, this paradise can be destroyed by adding additional products to satisfy additional customers. It is difficult to maintain production of a single product at a constant rate when demand for the others must also be satisfied. Production schedules for each of the products now must be created and managed. Now there will be changeovers that require both scheduling and people to manage them. Time will be lost to set-up. Quality will probably become more expensive, since with each changeover, the process has to be brought back into tolerance. Since there are additional products, each having its unique process requirements, additional process steps are likely to be required. Because it is much more difficult to match the capacities of each step of the process, it is very unlikely that the processes can be operated in unison. Inventories, too, will now be more difficult to manage. A greater variety of purchased items will need to be handled - in what is now an irregular pattern - to meet the production schedules. Work-in-process inventories can be expected to increase as inventories are built up to enable the many parts of the process to continue operating [...] Finished goods inventories will possibly increase because, while one product is being manufactured, stocks of other products have to be maintained to satisfy the demand for them. Customer priorities must be weighed against the priorities for smooth operation of the factory. As a consequence, the process is rarely in balance. In this factory almost nothing is perfectly predictable. So management costs are going to be much, much higher than those of the factory that manufactures only one product for only one customer. *(Stalk and Hout, 1990)*.

Womack et al. (1990), when describing Ford Model T factories, provide a good illustration of this point:

> Ford dramatically reduced set-up time by making machines that could do only one task at a time. Then his engineers perfected simple jigs and fixtures for holding the work piece in this dedicated machine. The unskilled workers could simply snap the piece in place and push a button or pull a lever for the machine perform the required task. This meant the machine could be loaded and unloaded by an employee with five minutes' training. (Indeed loading Ford's machines was exactly like assembling parts in the assembly line: The parts would fit only one way, and the worker just popped them on).
>
> In addition, because Ford made only one product, he could place his machines in a sequence so that each manufacturing step led immediately to the next. Many visitors to Highland Park felt that Ford's factory was really one vast machine with each production step tightly linked to the next. Because set-up times were reduced from minutes - or even hours - to seconds, Ford could get much higher volume from the same number of machines. Even more important, the engineers also found a way to machine many parts at once. The only penalty with this system was inflexibility. Changing these dedicated machines to do a new task was time consuming and expensive.[2] *(Womack et al., 1990)*.

Blackburn and Millen (1986), also referring to the findings of George Stalk in a previous study, report 'tremendous impact of increased products and processes variety on indirect costs or overhead which includes costs with schedulers, expediters, inventory trackers, space requirements, set-up personnel and so on. These people and systems have been introduced because of the complexity caused by multiple products, parts, process sequences and tasks.' In an example of a lift truck manufacturing plant in Japan, an increase of approximately 30 per cent in the overhead/unit cost was reported for every time the number of product families produced doubled (within a range between six and 20 families).

The disadvantage of excessive variability is also behind Skinner's (1974) concept of focused factory: 'A factory that focuses on a narrow product mix for a particular market niche will outperform the conventional plant, which attempts a broader mission. Because its equipment, supporting systems, and procedures can concentrate on a limited task for one set of customers, its cost and especially its overheads are likely to be lower than those of the conventional plant ... which ... attempts a complex, heterogeneous mixture of general and special purpose equipment, long and short run operations, high and low tolerances, new and old products, off-the-shelf items and customer specials, stable and changing designs, markets with a reliable forecast and unpredictable ones, seasonal sales, short and long lead times, and high and low skills'. According to Skinner, the fact that each of the contrasting features generally demands conflicting manufacturing tasks and hence different manufacturing policies is typically not well understood. The result, according to Skinner, is complexity, confusion, and worst of all a production organization which, because it is spun out in all directions, lacks focus and a 'doable' manufacturing task. Excess variability of outputs would not only jeopardize cost efficiency but, more broadly speaking, competitiveness.

One implicit assumption in the aforementioned points is the trade-off - a relationship in which any increase in variety implies a reduction in cost efficiency. The trade-off was formalized in the 1970s by Skinner (1974) although in recent years it has been questioned since some world class companies have managed to be successful in improving on all fronts. Slack (1990a) proposes a different view of the trade-offs between variety and cost efficiency. The author proposes that flexibility is the pivot which determines the relative state of, on the one hand the various types of variety, and on the other, the cost efficiency of the operation. Then, the more flexibility an operation has or develops, the closer the pivot gets to the cost-end of the seesaw. Therefore, more variety would be traded-off by less cost-effectiveness reduction. Slack also relates types of variety to types of flexibility which he considers appropriate to deal with. See Figure 3.1.

Variety types	Flexibility types
product variety	mix flexibility
new product introduction	product flexibility
output variation	volume flexibility
schedule/due date changes	delivery flexibility

Figure 3.1 Variety and corresponding flexibility types (Slack, 1990a)

Thus, according to Slack, flexibility is a way to achieve variability of outputs cost-effectively. The relationship variability-flexibility is further discussed in Chapters 4 and 5.

The reasons for product and parts proliferation

Some of the reasons for the companies' decisions to increase the product and parts proliferation can be identified in the literature. The first reason is related to traditional cost accountancy system procedures. According to Johnson and Kaplan (1987), the practice of using direct labor rates to allocate overhead costs to products causes, among other problems, the distortion of product costs and introduces unintended cross subsidies by shifting costs from less labor intensive products to more labor intensive products. That kind of practice would shift the variable overhead costs of products with relatively low direct labor content onto products which are more labor intensive. Costs, therefore, would be shifted from small-volume (per product), frequent set-up jobs onto long-running, infrequent set-up standard products. In this situation, the factory would, according to the authors, start to take on a broader product line (becoming a 'full-line producer') which means more low-volume products requiring frequent set-ups and so on. Thus, the mature, high-volume infrequent set-up, stable products become 'more costly' as the firm expands its product line and offers special features. The mature products this way subsidize the firm's product proliferation activities through a direct labor cost allocation system.

The second reason for product proliferation is the pressure from marketing and sales functions in order to have a broader or more complete range of products to offer to the customer (this issue is further discussed in the next section).

A third reason can still be identified and is related to parts rather than product proliferation. According to Neal and Leonard (1982), designers are expected to produce part designs which are simple, interchangeable and easy to manufacture. However, possibly caused by indecisive management pressure, lack of status, time, information or training, they attempt a design solution which is new and interesting but which tends to be non-standard and complex.

The strategic benefits of product variety

The benefits of variety are generally linked to the market. According to Womack et al. (1990), companies that have mastered lean design[3] are taking advantage of their strength in the market place, offering a wider variety of products and replacing them more frequently than the competitors. The authors suggest that this is one of the reasons for the successful performance of the Japanese auto industry across the world in the 1980's.

Given that the advantages of product variety are linked to the market, let us try to understand product lines management from the perspective of the Marketing function, by analyzing how a widely adopted Marketing textbook (Kotler, 1991) approaches the issue.

Kotler defines *product mix* as 'the set of all products and items that a particular seller offers for sale to the buyers'. A product mix may have a certain breadth, length, depth and consistency. The breadth refers to the number of product lines the company carries; the length, to its total number of items; and the depth, to the number of variants offered of each product in each line. Consistency of the product mix refers to how closely related the various product lines are in end use, production requirements, distribution channels, or other criteria. These four dimensions of the product mix would then support the company's strategic planners' decisions in defining the company's product strategy. The company, according to Kotler, can expand its business in four ways: adding new product lines or broadening its product mix; lengthening each product line with more products per line; deepening its product mix, with more variants per product; and altering (increasing or reducing) the product-line consistency, depending upon whether it wants to acquire a strong reputation in a single field or participate in several fields.

A *product line* is defined as 'a group of products that are closely related because they perform a similar function, are sold to the same customer groups, are marketed through the same channels, or make up a particular price range' (Kotler, 1991)

In order to decide about product line issues, according to Kotler, managers need two types of information. First they must know how well each item in the line sells and how much each profits. Second they must know how their product lines compare with competitors' product lines. A high concentration of sales in a few items is considered as 'line vulnerability' and these items should be closely monitored and controlled. On the other hand the manager could also consider dropping from the line the slow-selling items (items representing a low percentage of sales and profits). The importance that cost accounting practices have in this kind of decision and the risk of using inappropriate cost accountancy procedures were stressed in the last section.

In deciding about length, the issue would be influenced by company objectives. Companies seeking high market share and market growth will carry longer lines. They are less concerned when some items fail to contribute to profits. Companies that emphasize high profitability will carry shorter lines consisting of carefully chosen items. Kotler (1991) extends his discussion saying that product lines tend to lengthen over time and that not only does excess manufacturing capacity put pressure on the product line manager to develop new items but also the sales force pressures the management for a more complete product line to satisfy their customers. This 'line-stretching' can happen upwards (towards the more sophisticated end of the market), downwards (towards the less sophisticated end of the market) and two ways. There is also the possibility of line-filling, which is the line lengthening by adding more items within the present range of the line. Line modernization and product featuring are other reasons for introducing new products. According to Kotler, the two reasons for a company to eliminate products are when the product line includes 'deadwood' that is depressing profits and when the company is short of production capacity. Another way of generating product-line variety is through packaging. Physical products require packaging decisions to create such benefits as protection, economy, convenience, and promotion.

In terms of manufacturing operations, not always is the variety of different final products directly translated into additional parts, product routes and so on because many different products use common parts. On the other hand small differences in terms of marketing, such as packaging, can represent different processes, machines, and so on, translating into large variety for manufacturing.

It is somewhat preoccupying that such a broadly adopted and referenced marketing text book, in its 7th edition, published in 1991, still makes such little reference to manufacturing issues when discussing the company's product line-related decisions. Kotler (1991) virtually ignores the issues raised by a number of authors (Skinner, 1985; Hayes and Wheelwright, 1984; Hill, 1985; among others) regarding the need for a closer marketing-manufacturing relationship in deciding matters such as product line stretching or deepening. It does not seem to be advisable for companies nowadays to adopt such a top-down approach for strategy designing, leaving little space for the bottom-up, proactive role which the other functions (including manufacturing) should have. The consideration of the manufacturing current and potential abilities, for instance, viewed by a number of authors as crucial for good decision making on product line managing is not mentioned in the whole Chapter on 'Managing Product Lines, Brands and Packaging' of Kotler's (1991) book. Focusing is not included either, as a reason for product line breadth reduction, although a number of examples can be cited of companies which set up policies in order to reduce variety and increase focus. Variety is still justified by Kotler in terms of increasing excess capacity utilization, disregarding the now broadly accepted fact that utilizing a resource badly can be even worse for the company's competitiveness than not utilizing a resource (see Goldratt, 1988).

Despite these flaws, possibly still due to a lack of effective communication and understanding among different company functions, Kotler's approach to product variety was included in this review because it reflects the way a number of companies still deal with the problem. The intention of the review of Kotler's work is, therefore to highlight once again the problem represented by the inter-functional communication barriers within the organization.

Some conclusions about variability in manufacturing systems

At least two dimensions can be envisaged when analyzing variability of outputs: one is the actual variety of outputs which refers to the range of different products manufactured by the system, and the other one is the variation of the system's outputs during the time period.

The costs of an organization are in general very sensitive to the amount of variety. Excess variety generally causes an impact on indirect costs or overhead which includes costs with schedulers, expediters, inventory trackers, among other people, and with systems, introduced because of the complexity of multiple products, parts, process sequences and tasks. According to Slack (1990a), flexibility is a way to achieve variability of outputs cost-effectively.

The main causes of product proliferation, according to the literature, are the traditional cost accountancy system procedures, which favor the proliferation of products, and the pressure from marketing and sales functions which tend to attempt a broader or more complete range of products to offer to the customer. The benefits of variety are generally linked to offering more alternatives to the market.

Variety of outputs (both of products and of volumes) is regarded by a number of authors (Zelenovic, 1982, Goldhar and Jelinek, 1983; Stecke and Raman, 1986; Slack, 1989) as one of the main reasons for an organization to seek manufacturing flexibility. As present and future markets seem to show a growing segmentation, and, at the same time, the product life cycles tend to be shorter, the ability to produce a highly variable output have a tendency to become an increasingly important competitive feature for manufacturing companies. However, excessive or unnecessary variety should be avoided in order to help keep manufacturing focus and also because, as discussed earlier in this Chapter, variety invariably causes increasing costs and organizational disruption. A flexible system can, according to the literature, soften this negative effect of variety.

Notes

1 This is one of the reasons for the separate treatment of uncertainty and variability in the present research.
2 The Ford case is illustrative, although we have to consider that nowadays the car manufacturers are under a completely different technological paradigm. The Toyota Takaoka factory today, for instance, is able to change over in a few days from one type of vehicle to the next generation of products, while Highland Park was closed for months in 1927 when Ford switched from the Model T to the new Model A.
3 The wasteless agile design and development method that successful Japanese and western companies are using is based on strong leadership, emphasis on team-work, effective communication and simultaneous (as opposed to sequential) development of the various design and development phases; it has allowed such lean companies to shorten substantially the time and amount of resources spent in launching a new product.

4 Linking uncertainty, variability and flexibility: the management of unplanned change

Objective and summary

Chapter 4 is a theory building exercise. The objective is to attempt to make some progress towards the development of a conceptual framework aiming to help understand, analyze and manage the links between uncertainty, variability and flexibility in manufacturing systems.

In order to attain this objective, initially the relevant literature on the field is briefly reviewed. The literature review, the concepts discussed in Chapters 1, 2 and 3 and the field study (described in Appendix 2) are then used to support and guide an analytical process which will result in the identification of an important concept: that of management of unplanned change (of which uncertainty and variability are found to be only attributes) affecting manufacturing systems.

Five different types of unplanned change are identified and an approach to the management of unplanned change is proposed, including two complementary concepts: flexibility and unplanned change control. Unplanned change control is related to actions which are intended to avoid having to deal with the changes. Flexibility, on the other hand, is related to the decisions and actions which are aimed at dealing with the effects of the unplanned change left uncontrolled for some reason. Types of unplanned change control and types and dimensions of flexibility consistent with the proposed approach are developed and discussed.

The last part of the Chapter explains the role of the different types of manufacturing resources in determining both flexibility and unplanned change control. A new way of approaching the flexibility of the structural manufacturing resources is proposed. This new approach is based on the observation that the flexible structural resources always possess some level of redundancy in terms of their capability, capacity and/or utilization. The approach provides a new way of understanding the role of stocks in the achievement of flexibility in manufacturing system.

The analyses present in this Chapter draw heavily from the case studies described in detail in Appendix 2, to which the reader is encouraged to refer.

Introduction

Although a number of authors in the literature suggest that the environmental uncertainty and the variability of outputs are the main reasons for an organization to seek manufacturing flexibility, little empirically supported research work has been found which explores the mechanisms behind these relationships. In an attempt to fill this gap, the overall objective of this research is to understand and explore the relationships between variability of outputs, environmental uncertainty, and flexibility of manufacturing systems. The attainment of such an objective involves further exploration of propositions drawn from the literature which suggest the following general model:

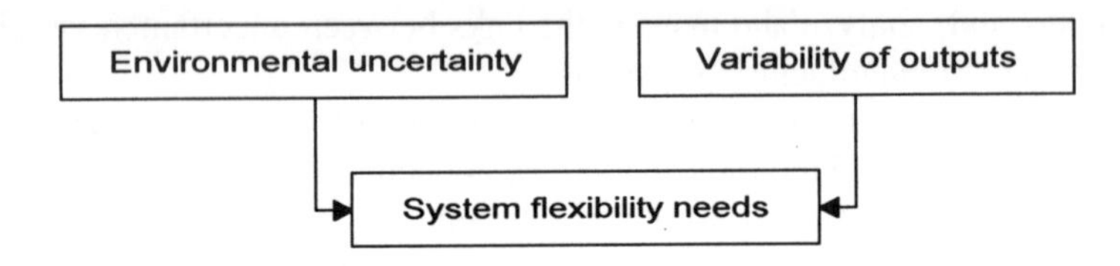

Figure 4.1 Reasons for flexibility according to the literature

No formal hypotheses as such will be established a priori. Rather, the major aim of this research is to build theory by constructing a model which reflects, organizes and possibly expands the perception of managers themselves regarding the aforementioned variables and their relationships. However, establishing some research questions (discussed in a later section) will help to determine the starting point for further analysis.

As a preliminary stage in describing the direction of this research, it is necessary to return to the literature in this and related fields.

The uncertainty - flexibility relationship

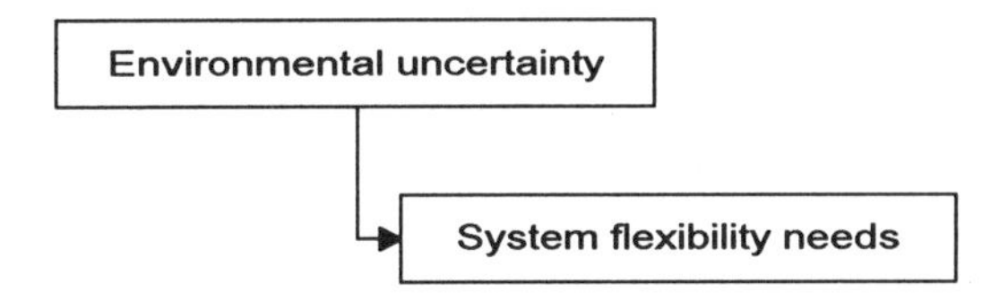

Swamidass and Newell (1987) develop a model incorporating the variables environmental uncertainty and manufacturing flexibility, tests it empirically and based on the results, states that 'an organization may find at least some help in coping with the high uncertainties imposed by the environment by increasing its manufacturing flexibility'. Gerwin (1986) argues that 'social systems facing uncertainty utilize flexibility as an adaptive response'; he further suggests that, since there are several kinds of uncertainty, there should be several kinds of corresponding flexibilities to cope with them. Gupta and Goyal (1989) suggest that flexible manufacturing systems can utilize flexibility as an adaptive response to unpredictable situations. Slack (1990a) also suggests that companies use flexibility to cope with short and long term uncertainties. Gerwin and Tarondeau (1989) take the analysis one step further, using Gerwin's (1986) classification, by suggesting links between particular types of flexibility and different types of uncertainty.

Atkinson (1984) argues that companies seem to be trying to develop more flexible manpower structures to be able to cope more efficiently with uncertainties regarding the supply of labor. Flexibility may also be developed as an insurance (Carter, 1986) against process short term uncertainty (Stecke and Raman, 1986). A more in depth discussion on the issues of uncertainty and flexibility can be found in Chapters 3 and 2 respectively.

The variability-flexibility relationship.

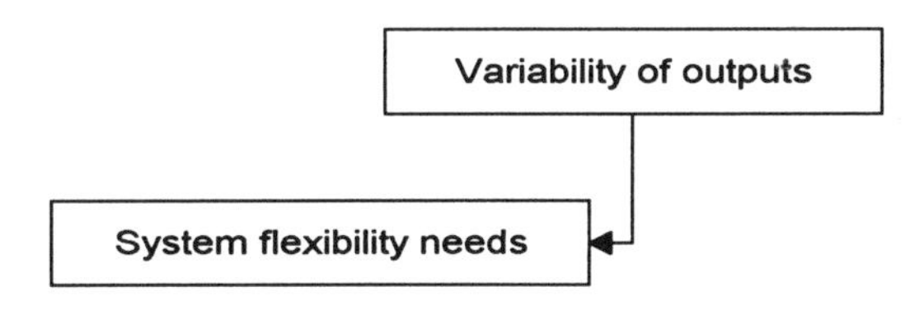

Variability together with uncertainty has formed the rationale for the operation's interest in flexibility. Flexibility would, according to Gupta and Goyal (1989), allow the organization to change its competitive strategy from economies of scale to economies of scope (Goldhar and Jelinek, 1983), as set-up time decreases and small batch production can be as economical as large scale manufacturing. Flexible manufacturing systems are important, according to Muramatsu et al. (1985), for companies to be able to adapt to severe changes in the market. Gerwin (1986), Kumar (1987), Chambers (1990), Frazelle (1986) and Stecke and Raman (1986) also argue that the need for flexibility is increasing because of the changing nature of competition, which, nowadays is based more than ever on the responsiveness of the companies to different customer requirements, shorter product life cycles and greater product proliferation. Slack (1990a) analyzes the links between types of variety and types of flexibility. For a further discussion on the concepts of variability of outputs and manufacturing flexibility, see Chapters 3 and 2 respectively.

The avoidance of the need to be flexible

Although the point is not explored as much as one might have supposed, considering its implications, some authors suggest that flexibility is not necessarily desirable in all circumstances, given that flexibility would never come cheap (see e.g. Slack, 1988). Slack (1991) claims that 'organizations should not make their lives unnecessarily difficult by generating the need for flexibility internally, in order to cope with bad design, poor communication, lack of focus, excessive routing complexity and year-end spurs'. Instead, they should try to eliminate the causes of such imperfections, by controlling the uncertainties and complexities involved in the process. This is in accordance with Slack's (1987) empirical findings according to which 'managers seek to limit the need to be flexible' by trying to compete on a non-flexible basis, adopting modular product design principles and by confining the need to be flexible to parts of the manufacturing system. With regard to the issue of controlling uncertainty, Thompson (1967) argues that 'organizations are open systems faced with uncertainty and ambiguity, yet require certainty and clarity to operate in a rational manner'. Managers of the organization's technical core would therefore attempt to reduce uncertainty so as to maintain operational objectives.

General comments on the literature

Although the existence of some kind of relationship between the three concepts - variability, uncertainty and flexibility - is recognized in the literature, further research is still required to provide both empirical support for these relationships and a greater understanding of the mechanisms driving them. If flexibility, for example, is the remedy for dealing with both variability and uncertainty, there may be an overall rationale behind this relationship, something that links both the concepts of uncertainty and variability. The same way, if it is true that managers tend to avoid having to be flexible, what are the means they use to do so?

There seems to be a need for an overall theory, an overall rationale behind the aforementioned three concepts. This theory would help managers explain, analyze and make decisions with regard to flexibility, taking into account all the relevant variables involved rather than just one or some, treated in isolation. It is not clear in the literature, for instance, whether flexibilities of the same kind should be applied in dealing with variability and uncertainty or different flexibility types are prescribed contingently.

There appears to be insufficient understanding not only of the relationships between factors, but also of the very way in which flexibility is understood and viewed in its contribution to manufacturing performance. This is evident from the number of papers which are still concerned with defining the concept and dimensions of manufacturing flexibility and trying to find physical analogies (such as the shock absorber model recently proposed by Slack, 1991) to explain it.

The present research is an attempt to understand and investigate the above mentioned mechanisms further in an attempt to possibly build theory: a theory which accommodates the most relevant variables involved in the decision process with regard to flexibility and the different and segmented views found so far in the literature.

Overall research objectives

The objective of the research described in this book is twofold: firstly, to try to answer the question: How do managers regard the relationship between environmental uncertainty, variability of outputs and manufacturing flexibility?; secondly, to build theory, attempting to conceive a model which reflects, organizes and possibly expands the perception of the managers in order to help them analyze and understand issues concerning the relationships between environmental uncertainty, variability of outputs and manufacturing flexibility[1]. In order to be able to achieve both objectives, a four year case-study-based research project was conducted with manufacturing companies in both Brazil and England. The main results of the research are presented in the next sections.

Refer to Appendixes 1 and 2 respectively for details regarding the methodology applied and the case studies. They are not present in the main body of this book in order that readers who are more interested in the conceptual aspects and in the results of the research can have a more direct reading. At the same time, those readers who are interested in the methodology and in details of the case studies can delve more deeply into these aspects by going through the Appendixes.

From uncertainty and variability to the concept of change

Although a number of authors have suggested that flexibility is needed in order to deal with the intrinsic uncertainties and the variability of outputs which are always present in manufacturing systems to some degree, it was noticed from the case studies that, at the level of analysis[2] adopted in this research, the managers generally attempted, during the interviews, to 'translate' the abstract terms uncertainty and variability into terms which were more meaningful and closer to their activities. For example, variability with regard to demand mix was translated into, or thought of, as frequent process *changeovers* between products; uncertainty regarding machine breakdowns was translated into unexpected *changes* in the availability of the machinery which could be used to perform the necessary tasks; variability with the product line was translated into *changes* in the set of tasks to be performed, from old ones to possibly novel ones; variability with demand volume was translated into *changes* in the utilization rates of the plant and the work volume to be done. It was also observed that, according to the managers' viewpoint, both the variability and the uncertainty affecting their operation are linked to the concept of *change*. Uncertainty and variability, then, are regarded as attributes of change. By analyzing the managers' answers, it is possible to attain a better understanding of their views with regard to the concept of change, which is relevant to the present research. The next section discusses the concept of change, drawing contributions from the literature and from the field research.

Change: definition and segmentation of the universe

When dealing with change in organizations, the literature makes an important distinction between two major types of change: the unplanned changes and the planned changes (Cummings and Huse, 1989; Lawrence et al., 1976).

The first type, unplanned changes, are changes which happen independently of the organization's determination but to which the organization has to adapt, e.g. an unexpected change in demand, a machine breakdown or a faulty supply. In this research, such changes will be called the *stimuli* acting on the system. Stimuli are thus defined here as the changes, either internal or external to the organization, which are perceived by the system's managers as relevant to the system's working and which happen independently of any conscious managerial decision.

The second type, planned changes, happen as a result of the organization's conscious managerial decisions which are taken in order to alter some aspect of the organization or its relationship with the environment. The implementation of a new technology aiming at quality improvements and programs to improve the level of commitment of people to the organization's goals are examples of the second type of changes.

Change types	Planned change
	Unplanned change or Stimuli

Figure 4.2 Change types

Most of the definitions found in the literature on organizational change refer to planned change. Wieland and Ullrich (1976) consider change as an organizational response made in anticipation of substantial environmental changes which, in turn, are associated with environmental discontinuities. The authors do not go further in defining 'environmental discontinuities'. Benne (1961) adopts Kurt Lewin's definition: change would occur when an imbalance occurs between the sum of the restraining forces (those forces striving to maintain the status quo in the organization) and the driving forces (those pushing for change) which constantly affect the organization. Lawrence et al. (1976) also emphasize planned change in their definition: change would be an alteration in the organization design or strategy or some other attempt to influence the organization's members to behave differently.

In the present context, because the research does not especially emphasize planned change, a broader definition of change will be adopted, which is a modified version of Cummings and Huse's (1989), which in turn was based on Lewin's. Change in the present context is defined as 'any modification, originated internally or externally to the organization, of those forces keeping a system's behavior stable and running without the need for any special decision or action by any of its elements'. Whenever a modification happens to one of these forces which calls for any decision or action, we consider that a change has occurred.

The two types of change, unplanned change (which is alternatively called *stimuli* in this research) and planned change, represent concepts which are not mutually exclusive. Dealing with some types of stimuli may call for planned change. 'Organizations can use planned change to more readily solve problems, to learn from experience, *to adapt to other changes* or to influence future change' (Cummings and Huse, 1989). Changes in the available technology, such as the development of MRP II[3] (manufacturing resources planning) systems in the 1970s, for instance, led the companies which decided to use it to take a number of decisions and actions in order to consciously change (planned change) aspects of the organization in order to prepare and adapt to the new technology. In the present research, we will be interested in discussing the stimuli-type of change and the way the organizations manage it. This is because stimuli is the type of change which, according to the literature, calls for flexibility in the manufacturing systems at the level of analysis which interests us. See Chapter 2 for a discussion on the objectives of flexibility.

Stimuli: nature and a proposition of taxonomy

As open systems, manufacturing organizations are continuously subject to the influence of stimuli originated from a series of internal and external sources, namely the process itself, labor, the suppliers, the customers, the corporate management, the other functions and the competitors.

Stimuli sources	Process Labor Suppliers Customers Society Corporate & Other Functions Competitors

Figure 4.3 Main stimuli sources affecting manufacturing systems

The stimuli dimensions or attributes.

Variability and uncertainty can be seen as attributes of the unplanned change or the stimuli-type changes. A particular stimulus can be more or less certain (or predictable) and more or less variable. However, during the field study, it was noticed in the discussion with the managers that variability appears to be too broad a concept to allow for an adequate analysis at the level adopted in the present research. Generally, variability had to be specified in more detail to be analyzed by the managers. The managers also mentioned, on a number of opportunities, examples of unplanned change types which they usually have to manage. Such examples can help in the search for a taxonomy of stimuli. The following section mentions some examples from the field study.

Types of stimuli found in the field work

Novelty The marketing function of a Brazilian heavy weaponry manufacturer (one of the case companies of the pilot field study), facing a military off-road and light vehicles sales drop in the late 1980's, decided to launch a new line of products - jeep-type light vehicles - in the consumer market. This decision was made in an attempt to utilize the plant's idle capacity. Such a change in the marketing strategy represented a completely novel set of stimuli to the manufacturing system, e.g. new quality requirements, new competitive criteria and new production volumes to which they would have to respond. *Novelty,* therefore, seems to be a relevant aspect or dimension of stimuli for the study of manufacturing flexibility. It relates to the degree of novelty in the situation brought up by the change. The more novel a situation is after a change happens, the more flexibility will probably be necessary for the system to be able to adapt effectively to the new reality.

Frequency A division of a British car manufacturer (field work's Company A) which manufactures engines faces changes in its demand mix for engine derivatives on every shift. Some of such changes are due to frequent and unexpected changes in the schedule of its internal customer, the vehicle assembly line. Others are an intrinsic part of Company A's business, which assembles vehicles to order. This requires the engine plant to produce approximately 60 per cent of the total number (78) of engine derivatives on each one week, resulting in frequent machine and assembly line changeovers. Some Japanese motorcycle manufacturers are another, and perhaps less trivial, example of frequency of change. They have a broad variety of products. Therefore, even with a very stable 'frozen' production plan period (which could give the impression of a situation of few changes), their operation functions face and have to respond to frequent changes because they must produce a multitude of products within a limited period using a limited amount of resources (Stalk and Hout, 1990). Frequency, which relates to the frequency in the occurrence of the change, thus seems to be another relevant dimension of the stimuli, for the purposes of the present research.

Certainty Company A's engine shop had a high degree of uncertainty regarding its demand changes. The work of the engine shop and the paint shop were based on the same master schedule. However, because of unexpected changes in the paint shop's schedule owing to technical problems, the engine shop had its demand changed frequently so as to match the actual outcomes of the paint shop. Probably because of lack of coordination between both units, the engine shop assembly line schedulers did not know in time what car body was coming out from the paint shop and therefore what engine types should be produced. They had to schedule the engine's assembly line under conditions of severe uncertainty and, thus, according to one of its managers, to master the art of 'fire-fighting', or reacting quickly. *Certainty,* therefore, is another relevant dimension of stimuli for the analysis of flexibility. It relates to how complete and accurate is the information which the system has about the changes - either present changes (something that has changed but the system has not acknowledged for some reason) or future changes (the predictability of the change).

Size A fourth dimension, which is complementary to the first three, can be logically identified: similarly to the dimension novelty, it relates to the degree of difference between the new situation brought up by the change and the situation before the change. However, a change may be large, but not novel, predictable (not uncertain) and not frequent. Company D, for instance, has a highly seasonal demand, which causes large changes in its demand volume from summer to winter. Although Company D's demand curve shape is very predictable and not novel, the demands in both seasons are substantially different and probably call for a different managerial response from the response demanded by the three first stimuli types. The fourth dimension thus relates to the *size* of the change.

Rate There is a fifth dimension of change which is relevant to this research. Change in demand volume is one of the main concerns of Company C's managers (see Appendix 2 for details). They have this concern not only because the changes are uncertain and large but mainly because the demand volume changes considerably in a very short period. One single large order can represent a considerable percentage of the annual production of the company. In order to fulfill the order, they may have to change their output rate considerably in a very short period. In the words of one of Company C's managers:

> Last week, for instance, an American buyer came to us and ordered 128000 shock absorbers. This represents 10% of our annual production... We will have to struggle to deliver them in the four month period we promised. (Company C manager)

Responding to this sort of 'steep slope' in the demand curve probably requires that the organization develop different abilities than those which would be required in order to respond to changes of the same magnitude (size) but which happen at smaller rates. The *rate* of the change seems therefore to be a fifth relevant dimension of change for the purposes of this research.

Figure 4.4 gives some examples taken from the four companies analyzed in the field work.

	Company A	Company B	Company C	Company D
Size	n/a	n/a	n/a	demand: seasonal with reasonably known general pattern
Novelty	technology: changes due to new emission control laws	technology: for the new line of products - fuel injection systems	technologies: concerning the high technological content products	n/a
Frequency	demand mix: products made to order which makes each and every vehicle different	machinery: frequent machine breakdowns due to unreliable and dated machinery	suppliers: frequently missing due dates (25% of the deliveries)	demand mix: a broad range of possible products to be made with scarce resources
Certainty	cell demand mix: changes in the engine assembly line program	supply: erratic supplies in terms of dates and quality	government policies: changes in the exchange rate policies of the government	demand mix: changes due to the diversified set of customers (from 140 countries)
Rate	technology: technology involved with diesel engines evolving too quickly	n/a	demand: large orders changing the demand curve dramatically in a short time period	n/a

Figure 4.4 Examples from the field study with regard to change types

Summarizing, based on the field work and on logical analysis, a taxonomy is proposed in order to analyze stimuli and their links with flexibility: there are five dimensions of stimuli which are relevant to the analysis of the manufacturing systems flexibility, at the level which interests us in the present research: the size, the novelty, the frequency, the certainty and the rate of the stimuli. Putting it in other words, the pattern of stimuli to which the manufacturing systems are exposed can vary in terms of its magnitude and dynamics. In terms of the magnitude of the stimuli, size and novelty are two relevant dimensions. In terms of the dynamics, frequency, certainty and rate are other relevant dimensions. Within the organization, each stimulus triggers a perception of the effects it will cause. The stimuli are perceived by managers as meaning either threats to or opportunities for the organization's competitive position. The management of the stimuli is an important part of the manager's job in the sense that it helps the potential threats to be minimized and the opportunities to be explored.

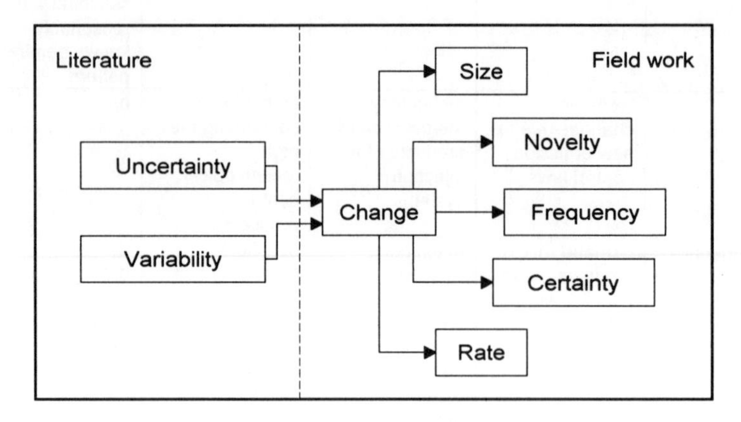

Figure 4.5 Stimuli-type change dimensions

Managers dealing with change

According to the literature, flexibility is needed in order to deal with uncertainties and variability of outputs in manufacturing systems, and to a certain extent, this suggestion is confirmed by the field work developed in the present research. However, it was noticed that the managers consistently approached the matter in a somewhat unexpected way. One of the most remarkable observations about the managers was their similar general approach to the management of stimuli. Invariably, two concepts came into the scene when the managers described the ways they usually deal with the stimuli-type of change. When the managers were asked, for instance, about the ways which they considered appropriate for dealing with uncertainty and variability, they frequently emphasized manners of attempting to eliminate or reduce the levels of uncertainty and variability of the changes which they would have to deal with. They would thus be seeking to avoid or reduce the need to be flexible. In other words, they would not only try to act after the fact, *responding* to the changes (by being flexible), but they would frequently prefer to act before the fact, endeavoring to *control* (in the sense of restraining or regulating) the uncertainty and variability of the unplanned changes which they would otherwise have to deal with. It is important at this point to clarify what is meant by *control* in this context. The term control generally includes some sort of feedback; however, when used in operations management literature, it frequently includes a broad array of different elements such as despatching, planning and scheduling. Control is a term which is in general loosely defined in the operations management literature. Different authors seem to consider control with different meanings. According to Schmenner (1990), there are four main functions that are reasonably identified with production control:

i) assigning jobs specifically and sequentially to each work center;
ii) monitoring the performance of actual production versus the schedule and informing management of the status of orders;
iii) taking action to remedy the unacceptable status of some jobs; and
iv) being an architect of information flow in the process.

Voss et al. (1985), addressing service operations management, argue that the term operations control relates to a series of aspects which include a wide range of different activities such as contacting customer, diagnosing customer problems, filtering customers, despatching and sequencing of jobs, selling, controlling information and invoicing. More specifically, Wild (1989) adopts the system's approach for control more strictly: 'control derives from the process of monitoring activities and the comparison of actual and intended states'. In the context of this book, the term control, when associated with change or one of its dimensions, means simply 'a means of restraining or regulating'[4]. There is not a commonly accepted meaning for the term control in the context of operations management. However, in order to avoid confusion with other definitions, when 'restraining or regulating change and its dimensions' is meant, the text will be explicit, i.e. by using terms such as change control, unplanned change control, stimuli control, uncertainty control and so on or by making the meaning clear from the context itself.

Examples of the use of unplanned change control and flexibility from the case studies

When arguing about appropriate ways of dealing with unexpected machine breakdowns, for instance, a number of managers answered that the ideal way is to improve preventive maintenance (to avoid the uncertain changes in machine availability, caused by the possible breakdown). With regard to those breakdowns which preventive maintenance could not avoid for some reason, the managers mentioned that the system should be able to take fast corrective action (e.g. by sourcing the necessary spare parts quickly and/or by re-routing the production flow), after the breakdown. In a similar way, a number of managers suggested the standardization, for instance, as a preferred manner of dealing with the variability of parts and products, and in doing so, avoiding the need to cope with such variability. For the cases in which the market really demanded product mix variability and standardization was impossible or inconvenient for some reason, they would then suggest, for instance, that being able to perform fast set-ups or developing labor multi-skills is important for coping with the variability of the product mix (see Appendix 2 for more examples).

	Case A	Case B	Case C	Case D
Control	**changes** caused by variability. **solution:** standardization	**changes** in the supply chain. **solution:** supplier development	**changes** in the supply chain. **solution**: coordination with suppliers	**changes** in the demand mix. **solution**: forecast sensitivity
Flexibility	**changes** caused by variability. **solution:** Labor multi-skills	**changes** in the supply chain. **solution:** rescheduling capability	**changes** in the supply chain. **solution:** buffer stocks	**changes** in the demand mix. **solution:** fast setups.

Figure 4.6 Examples from the field work with regard to the use of control and flexibility

The fact that the managers mentioned ways to reduce the need to be flexible was not completely unexpected, since it had already been suggested by Slack's (1987) empirical findings. What was unexpected was the emphasis placed by the managers on trying to keep the uncertainty and variability of the changes under control. In view of the findings of the field work, it is surprising that the literature has neglected this aspect which proved to be a major concern for the managers and which is actually complementary to flexibility in the management of unplanned change: the control of the changes.

Control, as considered here, relates to the set of decisions and actions taken in order to restrain or regulate the level of uncertainty and variability, *ex-ante* the changes which the system would otherwise have to deal with. It is important to notice that, in the sense considered here, stimuli control does not mean exclusively interfering *directly* with the source of the stimuli. Doing so is only one of the ways of exercising unplanned change control. Substituting a machine which frequently breaks down, and thus causes frequent unexpected changes, is an illustrative example of exercising control by acting directly upon the source of the stimuli. Nevertheless, acting on the source is not the only form of control identified in the field study. A decision can be taken consciously in order to make a work unit or a manufacturing operation less exposed to the stimuli. Sometimes, this is done by altering aspects of the operation itself, without interfering directly with the source of the stimuli involved.

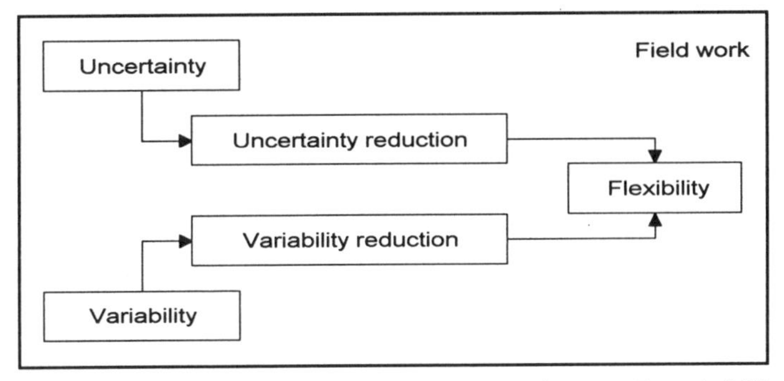

Figure 4.7 Managers emphasize uncertainty and variability reduction

An example is to focus a production unit on a specific range of products or on a specific task[5]. Company A, for instance, has its engine plant organized in manufacturing cells[6]. One of them is dedicated to machining only two basic types of engine blocks. The operators therefore do not have to perform frequent machine changeovers in this cell. By focusing the cell on a specific task, the plant manager restrains the amount of change which the cell 'perceives', although not interfering directly with the source of the changes which is possibly the demand mix.

The management of change: how the literature addresses it.

There is an extensive literature under the heading 'management of change', generally by researchers on organizational behavior. Their approach strongly emphasizes the management of planned change rather than stimuli. The question they endeavor to answer is basically 'how to change the organization effectively'. The management of stimuli is, in a way, neglected. The literature on production operations management usually deals with the issue of managing unplanned change under a number of different headings. One of them, which is evidently related to stimuli-type changes, is 'manufacturing flexibility' ('the ability to respond to changing circumstances', according to Mandelbaum, 1978). Although very valuable contributions are found in the manufacturing flexibility literature (Browne et al., 1984; Mandelbaum, 1978; Buzzacott, 1982; Zelenovic, 1982, among others), few (Slack, 1990a; Gerwin, 1986; Swamidass and Newell, 1987) attempt to understand, identify, classify and relate reasons for being flexible (the 'changing circumstances', or, according to the terminology used here, the 'stimuli') with different types of flexibility. They argue that flexibility is necessary in order to deal with uncertainty and variability, but, since their emphasis is on flexibility, they do not explore[7] the fact that uncertainty and variability can also be dealt with by *controlling* them.

Thompson (1967), on the other hand, worked on the idea of the manager's need to control uncertainties but, at least for this context, did not sufficiently explore the need to deal with the uncertain stimuli which were left uncontrolled. Gerwin and Tarondeau (1982) propose the adoption of flexible technology as an addition to Thompson's strategies for controlling uncertainty but they concentrate their analysis on the technological resources and on the long term uncertainties. They have not gone very far in actually discussing how the complementarity control/flexibility would work either.

The control of stimuli is also treated, although not explicitly, under a number of research headings. Manufacturing focusing, vertical integration and make-or-buy decisions (which , as will be explored in Chapter 5, can be seen as particular cases of stimuli control), for example, also have a rich research literature, but each of them is invariably treated in isolation.

Based on the previous evidence from the field study, an alternative approach to the ones found in the literature is proposed here. According to the proposed approach, there are two distinct ways used by managers in order to handle unplanned change in manufacturing systems:

i) by controlling the unplanned change and therefore by interfering either directly with, or with the way the manufacturing system perceives, the size, novelty, frequency, certainty and/or rate of the changes, before the changes occur.

ii) by dealing with the effects of the stimuli by being flexible, which is the ability to respond effectively to the changing circumstances, after the changes.

Figure 4.8 represents the reasoning of this proposed approach.

To summarize, according to the literature, variety and uncertainty are the main reasons companies develop manufacturing flexibility (top box in Figure 4.8). From the field work, there was evidence that uncertainty and variety always referred to change and that a more appropriate way of classifying change for the purposes of this research was in five dimensions: size, novelty, frequency, certainty and rate (bottom box in Figure 4.8). Also from the field work, there was evidence that, in addition to their concern with the need to respond to change, the managers frequently emphasized their concern about the possibility of reducing the levels of uncertainty and variety with which they have to deal (second box from the bottom in Figure 4.8).

The concurrence of these aspects results in the proposed alternative approach, represented by the second box from the top (in Figure 4.8): unplanned change has five main dimensions - size, novelty, frequency, certainty and rate. To manage these unplanned change dimensions, managers adopt a mixed approach contingently - preferably they seek to control the occurrence of change (before the change). They then develop flexibility in order to be able to deal with the effects of the unplanned change (after the change) which were left uncontrolled.

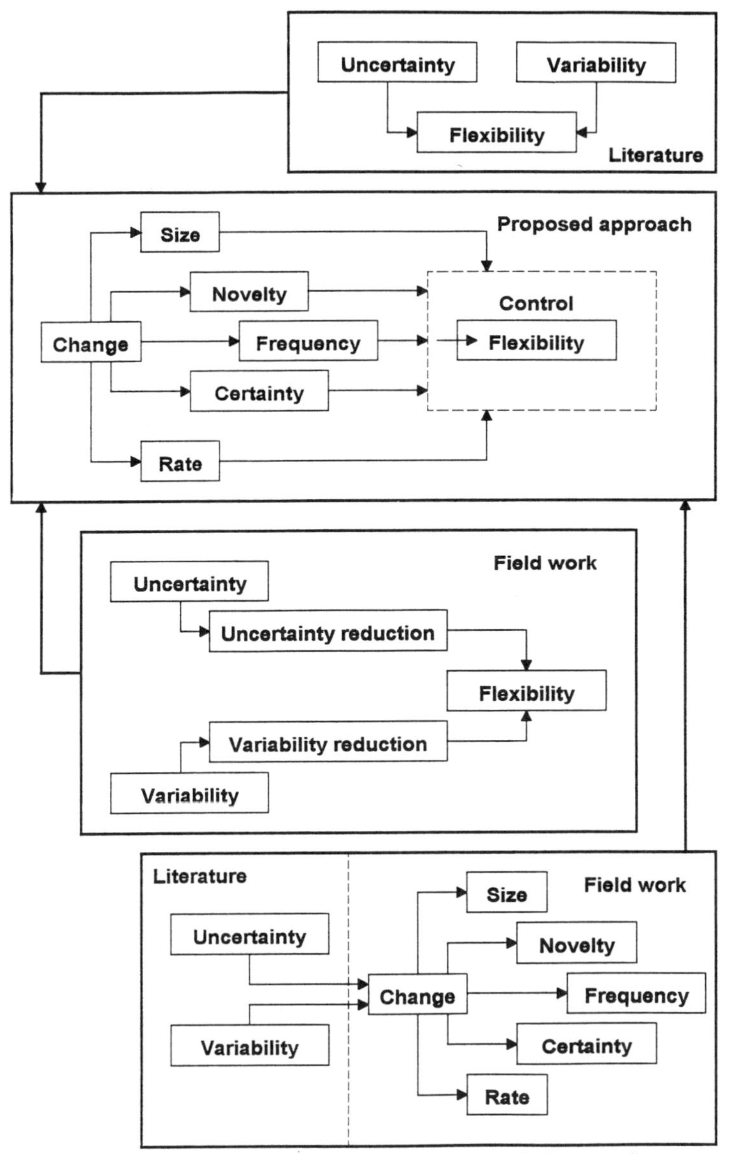

Figure 4.8 Schematic development of the proposed alternative approach

Control: managing the influx of the stimuli

There are several methods used by the managers of the case studies to control their perceived *influx* of *stimuli* (the level at which the organization perceives and is influenced by the *stimuli*). Some of them relate to interfering directly with the sources of the stimuli whilst others relate to interfering with the way the system is affected or chooses to be affected by the stimuli. Some of the ways which can be used in order to control the stimuli are described by the examples. The examples are drawn from the field study (see Appendix 2 for details).

Examples of unplanned change control types from the field work

Monitoring and forecasting Facing a turbulent environment in terms of industrial relations, Company C closely monitors the trends of the behavior of the labor unions in Brazil, in order to avoid being taken by surprise, for instance, by a labor strike. In doing so, Company C is trying to increase the predictability or reduce the uncertainty of some of its stimuli.

They also adopt 'monitoring' as a way to keep up with the new process and product-related technological developments. With this aim two offices were established by Company C, one in the United States and one in Germany. In this manner, they are attempting to reduce the novelty of the stimuli which they would have to deal with if they only took notice of a new technology when it had already been completely developed. Thus Company C uses *Monitoring and forecasting* as ways to control some of the dimensions of their stimuli.

Coordinating and integrating Company A's engine manufacturing shop reduced its short-term demand uncertainty by establishing on-line computer links in order to coordinate the engine shop with the paint shop. With on-line information, the engine shop now has accurate and timely information about the car bodies which are coming out of the paint shop and therefore they have better information about the next few hours' demand for engine derivatives. This achievement allowed them to schedule the assembly line more effectively, under less uncertainty. Another example of reduction of uncertainty by coordination is the notorious change that has happened in recent years in the customer-supplier relationship (of which the relationship between Toyota and its suppliers is a representative example); it has been a change from confrontation to cooperation and integration (Womack et al., 1990). The reduction of the supplier base, the tendency to establish long term stable contracts, with strong emphasis on personal contacts, are mechanisms used by some organizations to increase the integration and control over the changes with their supply. Considering internal suppliers (sectors of the manufacturing systems which supply other sectors), another example of coordination is the use of pull systems[8] in order to coordinate downstream demand with upstream operations, using visual techniques such as Kanban cards[9]. Upstream vertical integration by acquisition of suppliers is another possible way of integrating and therefore increasing control over the changes related to supply. This approach has been largely utilized by Company C, which, over a period of years, has bought out a number of either uncertain or unreliable supplier companies[10]. *Coordinating and integrating* therefore are actions which are used in order to control the stimuli to which they are exposed. They can primarily influence the certainty of the change.

Focusing and confining Company A's engine shop adopts the focused manufacturing approach[11], organizing its machine shop in work units or cells. Company A's cells are generally set up to perform a limited range of parts. The cell which machines the engine blocks, for instance, uses automated transfer lines in order to perform only a few slightly different engine block types. On the other hand, another cell is manned with multi-skilled workers and equipped with computer numerically controlled - CNC - machines to perform a multitude of aluminium and steel engine components with considerably different characteristics. This way the need to be flexible is confined to one production unit or cell whilst the rest of the machining cells work only on a limited range of parts each. With the focused approach, depending on what sort of task the system decides to focus on, the size, novelty, frequency and/or certainty of the stimuli which is perceived by the system or part of the system, can be altered. If the chosen task is to produce a limited product range, when a hypothetical customer's demand pattern changes and he orders a completely different product, the company may then opt not to attend to it. This way, by focusing, the novelty of the change the system has to deal with is restricted. Another way of focusing would be, for instance, on serving only large orders, influencing the frequency of the system's machine changeovers. In contrast, the focus can be on flexibility, where organizations choose to focus the operation on manufacturing a large variety of products; the organization consequently invests in employee skills, process equipment and systems, which should then support the needs for flexibility. In this case, one way to exercise control over the stimuli, which the system as a whole perceives, is to confine the need to cope with substantial changes to a few flexible work units (see Appendix 2, case A, for a detailed example). This way, the amount of change which the rest of the operation has to deal with is controlled. In this sense, *Focusing and confining* were therefore identified as means used with the aim of controlling stimuli.

Delegating and subcontracting According to one manager of Company A, gradually some car manufacturers, including Company A itself, seem to be increasingly delegating, to suppliers or expert companies, the task of designing parts and components of their products. They are giving some of the suppliers only the design requirements and broad functional specifications about interfacing components instead of giving them detailed drawings and specification, as they used to do. This is one way in which these companies are limiting the amount of change, mainly in terms of the novelty and rate which they have to deal with, regarding product technology and design. Company A, for instance, had always designed its own diesel engines. However, in recent years they made the decision of subcontracting a European expert firm to design them, mainly because the technology involved with Diesel engine design was changing substantially (novelty) and at a very fast rate (among other reasons, because of new regulations with regard to emissions control). They considered that it would be more convenient for the organization not to try to keep up with the technology changes by using only internal design expertise. By *Delegating and subcontracting*, which relate to delegating to a contractor the need to cope with some of the changes, companies can control the stimuli they are exposed to.

Hedging and substituting Company B, facing a problem of erratic supplies, decided to run programs on supplier base reduction and supplier development. However, while the suppliers are still below the desired levels of reliability, the company has chosen to keep some of the standard components supplied by a number of sources rather than one or a few sources. This decision aims at hedging against the short term uncertain delivery the suppliers are still providing. By having a number of suppliers, Company B hedges against the uncertainty of one or some unreliable suppliers. If a company is relying on just one erratic supplier, it is probably more vulnerable to the undesired changes which the supplier can possibly cause. Although hedging is in a way contradictory to the general tendency of reducing the number of suppliers and developing a closer relationship with them, there may well be short term situations in which the organizations consider that hedging is a convenient way to control their uncertainties with regard to supply.

One of the most evident ways to limit the stimuli levels which an organization has to deal with is by substituting the source of the change, replacing it with a less 'changeable' one. If a supplier is consistently unreliable, for instance, frequently causing changes in the system's schedule by faulty deliveries, a company can reduce the occurrence of these changes by substituting the supplier, replacing it with a more reliable, certain one. The same applies to an unreliable piece of equipment which frequently breaks down (influencing the change frequency) and to a worker who is not dependable. *Hedging and substituting* are therefore also included among the ways in which organizations can control stimuli.

Negotiating, advertising and promoting Company D's manufacturing plant is running a program of parts standardization aiming at reducing the variety of parts which they have to manufacture. Such an effort involves negotiation with the plant's internal customer: the marketing function. By negotiating, the plant is trying to reduce the amount of change it has to cope with. Negotiating consists of an attempt to interfere directly with the customer (either internal or external) in order to reduce the changes she/he can possibly demand. Another illustrative example of negotiation is what happens with the firms which use Kanban systems (such as Toyota). Given that such firms need a stable environment in order to operate effectively, they generally 'freeze' their master plan for a considerable period of time in advance (Stalk and Hout, 1990). This aims at controlling the uncertainty and frequency of the short term demand changes. The management of this sort of change control also requires negotiation with the customers, be they internal or external. By negotiation, the demand curve shape is altered so that the system has to deal with less uncertain, smaller, less novel, less frequent or less drastic changes. Another way to interfere with the demand curve shape is by advertising, trying to influence customers to consume certain types of products or to induce determined patterns of custom which can also interfere with the frequency, rate and size of the future demand changes. Promotions and advertising campaigns are usual ways to stimulate off peak demand in order to level, or in other words, reduce demand change size and rate over time. *Negotiating, advertising and promoting* are therefore ways in which companies can control their stimuli.

Maintaining, updating and training Most of the managers interviewed during the field study mentioned preventive maintenance as a desirable way to handle machine breakdowns. A well maintained machine would be less subject to changes in its availability, caused by possible breakdowns. Maintaining the resources would thus be one way to reduce possible undesirable changes caused by equipment breakdowns, with regard to frequency and size. The idea of maintenance, however, is not only suitable for structural resources, such as machines. The maintenance of payment schemes and systems, in order to assure that they are updated and appropriate, and the maintenance of computer systems records to ensure data integrity are other ways to control possible future occurrence of severe changes (by reducing the possible size, rate, frequency and uncertainty of the change). Instances of such changes would be a possible unexpected, disrupting industrial dispute, or a late acknowledgment about some relevant inaccurate information in the computer records such as inventory quantities. With regard to human resources, one of the ways managers reduce the uncertainty of people's behavior is by training them in order to standardize procedures and increase their awareness of the importance of their activity and its impact on the overall performance of the operation. Four out of six Company B managers, all of them concerned with the uncertainties regarding the middle management's behavior under a major change which the company is to face, said that training was the most appropriate way of reducing the uncertainty and increasing the predictability in that respect. Therefore, *Maintaining, updating and training* are also ways in which companies can control their stimuli.

Summarizing, seven general types of control of stimuli-type changes were identified during the interviews. Figure 4.9 lists them and also shows where the identified control types fit into the proposed approach, developed in previous sections.

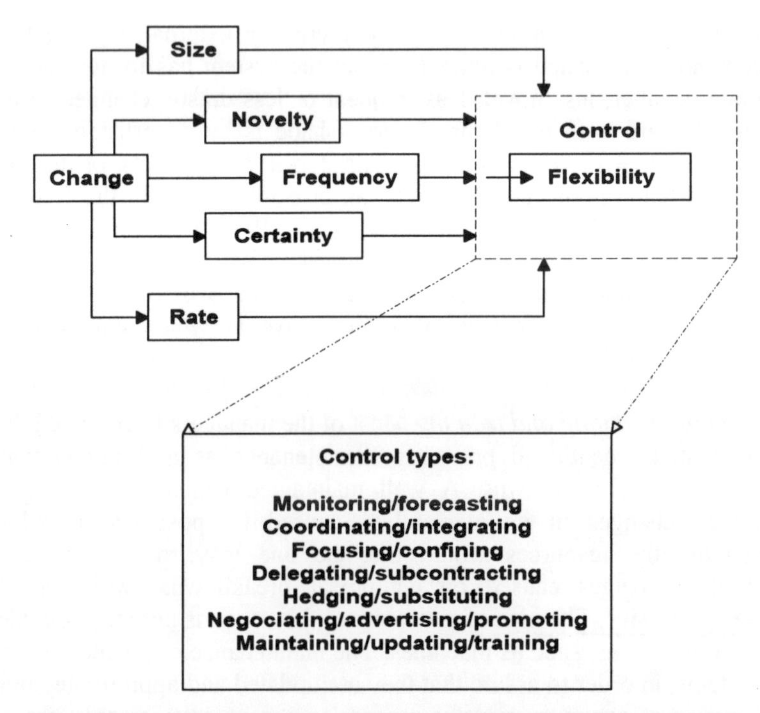

Figure 4.9 Detailing of the alternative approach: unplanned change control types identified in the field work

Flexibility: dealing with the effects of the stimuli

When studying the flexibility of manufacturing operations at the level of analysis which interests us in this book, we are primarily concentrating on the flexibility of the manufacturing system (the set of manufacturing resources), or in other words, the ability of the manufacturing system as a whole to respond to its stimuli. At this level of analysis, the flexibility of particular resources are only means to contribute to the achievement of the systems flexibility. This is also the most appropriate level of analysis if we intend to be able to understand the ways the manufacturing system can actually help the organization compete, bearing in mind the strategic role of the manufacturing function. In this sense, we assume that the performance of the whole system is more relevant for the organization than the performance of the particular resources, and therefore in the present discussion the particular resources will not be analyzed in isolation or in detail, but always as parts of a greater system.

Similarly, the decisions which are made beyond or at a higher level than the manufacturing operations management's level will not be emphasized here. Some authors, for instance, define expansion flexibility (Browne et al., 1983), as one of the manufacturing flexibility types. Although decisions regarding manufacturing unit expansion, through investments, acquisitions or other means, concern the manufacturing function, they are generally made beyond the level of decision of the manufacturing system. They are decisions generally made at the corporate or business level. Here, for the sake of maintaining the focus of this research, Browne et al.'s (1983) expansion flexibility, for instance, and other flexibility types will not be considered as manufacturing system's flexibility.

The consideration of flexibility here assumes a given core technology which encompasses the bulk of machinery, equipment and facilities which the manufacturing system already possesses and which, in general, cannot be substantially altered by decisions made at the operational level.

There are several classifications of manufacturing flexibility in the literature (see Chapter 2 for details). Some of them mix different levels of analysis. Others (such as Mandelbaum's action and state flexibilities) are overly broad and although valuable in the effort of conceptualizing flexibility, are of little practical use for the analysis of the manufacturing operations. At the manufacturing systems level, Slack's (1989) classification seems to be one of the most consistent. The author suggests that four types and two dimensions of manufacturing flexibility can be identified at the manufacturing system operation level: new product flexibility (related to the system's ability of introducing different products or modifying existing ones), mix flexibility (related to the system's ability of manufacturing a broad range of products within a given period of time), volume flexibility (related to the system's ability to change its aggregated level of output), and delivery flexibility (related to the ability of the system to change delivery dates). The two manufacturing flexibility dimensions defined by Slack are: range flexibility, the total envelope of capability or range of states which the operations system is capable of achieving, and response flexibility, the ease, in terms of cost or time, with which changes can be made within the capability envelope. Slack's classification was used in the interviews performed at the field work stage of this research.

Slack's four types and two dimensions were generally considered by the managers as valuable and consistent with their needs, at least with regard to changes with the system's demand. The managers usually understood the four types and two dimensions with ease and they were able to assess the performance of their operations in terms of each of them and identify the ones which they regarded as competitive priorities, recognizing the importance of such classification in terms of allowing the managers to establish priority actions and focus. In fact, logically, the system's demand can change in terms of its four main attributes of specification, mix, volume and delivery dates, which would be dealt with, respectively, by new product, mix, volume and delivery flexibilities. However, Slack's classification was not seen as sufficiently comprehensive at least for the level of analysis addressed in this research. The field study results suggest that, when analyzing change comprehensively, there is a need to define a complementary type of system's flexibility, which may bear some similarity to Mandelbaum's (1978) state flexibility[12]. A fifth type of system flexibility is then proposed in order to complement the four types proposed by Slack (1989). The fifth system flexibility type is related to the *robustness* of the manufacturing system, considered here as the ability of the system to overcome unplanned changes either in the process (such as labor absenteeism or machine breakdowns) or in its input side (such as faulty deliveries). Here, it will be called *system robustness* flexibility.

The need for a fifth systems flexibility type comes from the field study observation that even a system with high levels of performance in Slack's four flexibility types can lack flexibility to handle some of the changes which may affect the process or the input supply. A production unit could, hypothetically, have excess capacity (allowing for volume changes) and short set up times (allowing for fast mix changes), it could be very capable (being able to manufacture a large range of parts) and still it could have a machine which is the only one of its kind available in the unit, a machine which is the only one able to perform certain tasks. If this machine breaks down, for instance, the system's performance can be severely affected unless some sort of *system robustness flexibility* is present (such as a buffer stock after the machine, a responsive corrective maintenance system or an efficient outsourcing system, able to outsource the parts which otherwise would have been made by the broken machine). This was evident in Company B, which emphasized this sort of flexibility because their dated equipment was not very reliable (see Appendix 2, case B, for details).

In an attempt to explore further the concept of system's robustness flexibility, we can also think of this type of flexibility in terms of the two dimensions: range and response. The range dimension refers to how big the change or the disruption suffered by the system can be before its performance is significantly affected. The response dimension refers to how quickly, easily and cheaply the regular operation can be reestablished, once a disruptive change has occurred.

System's robustness flexibility is a way to achieve system's reliability by other means than by increasing the reliability of the individual resources. In other words, if a system works on the reliability of its individual resources, it would be exercising control rather than flexibility, because the intention is to avoid the occurrence, *ex-ante* the change. On the other hand, when a system develops system robustness flexibility, it is becoming prepared to deal with the changes, *ex-post* the occurrence of the change. Both approaches aim at increasing the overall reliability of the system.

Summarizing, from the evidence of the field work, it is proposed here that five types of system flexibility are relevant to the analysis of the manufacturing systems at the level analyzed in this research: new product flexibility, mix flexibility, volume flexibility, delivery flexibility - the first four from Slack's (1989) model - and systems robustness flexibility. These five flexibility types can be seen as having two relevant dimensions: range and response.

A correlation can be logically established between the types of change - system input-related changes, process-related changes or output-related changes - and the types of system flexibility - new product, mix, volume, delivery and system robustness. Changes relating to the output side of the system or to the system demand, that is, new products (or product changes), product mix, overall demand level and delivery dates, are mainly (although not exclusively) associated respectively with the aforementioned first four types of system flexibility - new product, mix, volume and delivery. Changes related to the input side and to the process elements (which can also be seen as inputs, as long as the system is analyzed with a long term perspective), which generally represent risk of disruption for the transformation process, are in turn primarily related to the fifth type of system flexibility - the system robustness flexibility.

In addition, there seems to be a correlation between the five stimuli dimensions of size, novelty, frequency, certainty and rate and the two flexibility dimensions of range and response. Size and novelty relate to the breadth of the change, to how different the new situation is after the change. Therefore, it is necessary that the resource or the set of resources involved with handling the change have the ability to assume a very different state (in order to deal with the size of the change) or to assume a large number of states (in order to increase the probability that one of them will match the novelty represented by the post-change or during-change situation). This suggests that changed size and novelty are related to range flexibility rather than to response flexibility.

Frequency, certainty and rate, on the other hand, relate to the dynamics of the change process. The faster and the more frequent and uncertain (unpredictable or unknown) the changes are, the more dynamic is the environment and the shorter is the response time required from the resource or set of resources, because these changes happen either unexpectedly, frequently or quickly. In other words, as the changes become faster and more uncertain and frequent, a greater response flexibility would be required. Figure 4.10 represents the five types and two dimensions of flexibility proposed and also shows how they fit into the general proposed approach developed in previous sections. Drawing from the different types of control and flexibility proposed, let us try to analyze and explore further the concepts of control and flexibility.

Unplanned change control and flexibility: exploring the concepts

The manufacturing system is a configuration of interacting individual resources which can be classified into:

i) technological resources - the facilities and technology or the hardware side of the manufacturing system.
ii) human resources - people in the manufacturing system.
iii) infrastructural resources - the systems, relationships and information couplings which bind the operation together.

The next section is an attempt to determine which, among the three basic types of manufacturing resources, are the ones which play the dominant roles in the achievement of the seven types of control.

Control and the manufacturing resources

Monitoring and forecasting There are related primarily to forecasting and information systems, although a company can have some people with expertise or experience in forecasting (see Appendix 2, case C, for an example).

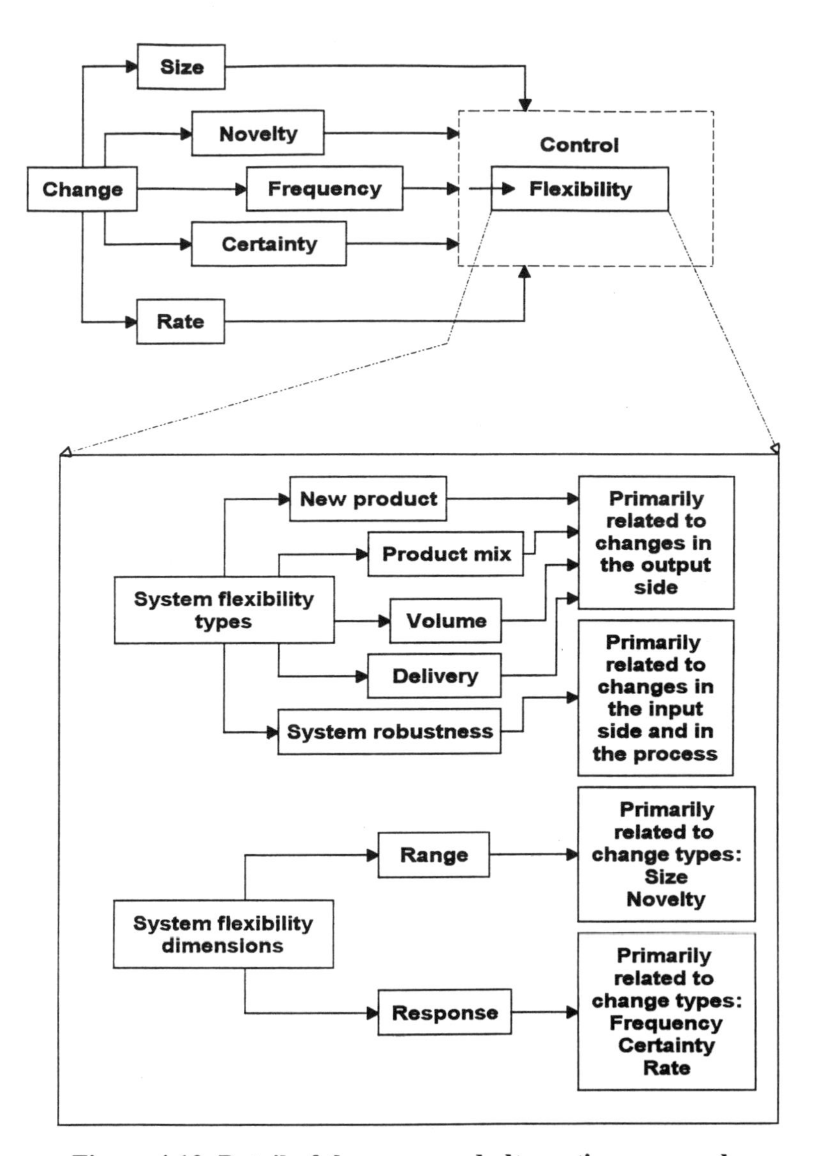

Figure 4.10 Detail of the proposed alternative approach: system flexibility types and dimensions

Coordinating and integrating In terms of the external suppliers, they are related to the supply management or to the information systems, depending on the type of coordination and integration the organization chooses to adopt. In terms of internal suppliers, it depends basically on the planning and control systems of the organization, which include inter-functional communication, forms of work organization and so on (see Appendix 2, case A, for an example).

Focusing and confining These depend on the organizational and work structuring which determine the work units and their tasks (see Appendix 2, case A, for an example).

Delegating and subcontracting They depend primarily on the make-or-buy policies and on the sourcing systems (see Appendix 2, case A, for an example).

Hedging and substituting In terms of supply and suppliers, they depend on the sourcing system and on policies regarding the relationship with the suppliers. In terms of substituting machines or people for less 'changeable' ones, they depend on technological and human resources and on the system and policies which are responsible for deciding for the substitution (see Appendix 2, case B, for an example).

Negotiating, advertising and promoting They depend basically on the systems and policies which determine the relationship with the customer (see Appendix 2, case D, for an example).

Maintaining, updating and training Here the idea is that reliability is built on the resources, making people, machines and systems less 'changeable'. They depend therefore on the three types of resources and on the systems which actually build up the resource reliability, by training people, maintaining machines and updating systems (see Appendix 2, case B, for an example).

Except for substituting resources for less changeable ones (hedging and substituting) and for 'building reliability' (maintaining, updating and training) on them, both being controls which involve primarily structural resources, the other types of control relate primarily to infrastructural resources. Infrastructural resources therefore seem to play a major role in developing ways to control stimuli in manufacturing systems.

Figure 4.11 summarizes the main contribution that infrastructural resources can give to the seven control types.

Infrastructural resource key players in the achievement of the different types of stimuli control		
Monitoring and forecasting	Achieved through	information and forecasting systems environmental and internal monitoring
Coordinating and integrating		vertical integration, MPC systems, supply chain management, organizational systems, inter-functional communication
Focusing and confining		organization structure, MPC systems, strategic planning
Delegating and subcontracting		make or buy policies, outsourcing, vertical disintegration policies
Hedging and substituting		supply chain policies, procurement systems investment policies
Negotiating, advertising and promoting		marketing/manufacturing interface marketing policies
Maintaining, updating and training		maintenance systems, training policies, information systems, industrial relations policies

Figure 4.11 Infrastructural resources: key players in the achievement of stimuli control

The structural resources have to be reliable, to be 'controlled'. More reliable resources will give fewer unexpected stimuli for the system to deal with.

Structural resources contribution to the achievement of the different types of stimuli control		
Substituting	increase resource	new more reliable machines and people
Maintaining/updating/ training	RELIABILITY through	'building up' reliability on machines and people

Figure 4.12 Structural resources contribution to the achievement of stimuli control

A similar exercise can be done with the relationships between the three types of manufacturing resources and the five types of manufacturing flexibility, drawing examples from the field study.

Robustness flexibility To deal with a faulty supply, for instance, Company A's engine plant developed a good ability to reschedule its production, giving priority to orders for which the components are already available. In this particular case, the rescheduling is mostly done by people. Another way of dealing with faulty supplies is to keep safety buffer stocks of raw material and components, such as the 'strategic' stocks of components built up by Company B, to prevent against possible problems with delivery.

To deal with the changes which affect the availability of process elements such as machine breakdowns and labor, the trivial way is to have excess capacity of the same resource. Company B, for instance, has excess process capacity in the Zamac injection shop, where the machines are not considered to be reliable and all the case companies keep excess people to cover absentees. In this regard, a non bottleneck machine, is, by definition, a machine which has capacity in excess. Even if it breaks down, provided that the time to get it running again does not exceed its idle time in the period, the system's working will probably not be substantially disrupted. If the time to make the machine up and running again exceeds its normal idle time and there is a risk that a subsequent bottleneck will starve (stop running because of lack of input material), one way to increase the system robustness flexibility is to build up some safety buffer stock - in general a time rather than a stock buffer[13] in front of the bottleneck resource. The '*bottlenecked*' Company C builds up stocks of iron powder every year in October in order to make sure that it continues running even if problems crop up with a North American supplier which is vulnerable to problems with transport during the American winter season.

An alternative way for a manufacturing system to be robust against breakdowns is, for example, to have a number of non-bottleneck machines which are able to perform each other's jobs (some extra equipment capability and capacity - the strategy used by Company B in its Zamac injection shop). In case of bottleneck machines, which have no excess capacity by definition, another way to increase system robustness flexibility is to have the capability to subcontract internal or external suppliers in order to do the broken down bottleneck machine's job. Company A uses this policy, keeping records of possible substitutes for its bottleneck-machines in and outside the corporation to which Company A belongs.

In terms of labor absenteeism, two aspects need to be considered, according to the observations from the field study: the availability of quantity of people and the availability of the right skills. When a worker is absent, someone else has to do his job. If the skills of the absentee are standard or non specialized, some excess capacity can make it up. Company A's engine assembly line keeps a certain amount of people in excess (some extra labor capacity) to cover for the 'normal' absenteeism rate which is assumed to be 3 per cent. Company A still have another problem, though. Not all the workers in the assembly line can do all the assembly tasks. Therefore, although they know the average daily number of absentees, who or what kind of skills will be absent in the next shift is a variable which is much more difficult to predict. In order to increase the system robustness flexibility with regard to absenteeism, Company A has improvement programs running which aim at training people in order to develop multi-skills (providing some excess labor capability) to allow for the labor transferability between tasks. This way, the line management increase the probability that, not only have they the right number of people to run the line every day but also that they have the right set of skills.

Summarizing, companies pursue the achievement of system's robustness flexibility in a number of ways by using technological, human and infrastructural resources. In this sense, apparently, there is not one type of resource which plays a major role.

Product flexibility Product flexibility seems to be dependent on a number of aspects: the human and technological resources must possess a 'reserve' of capability (see Appendix 2, case A) in order to be able to deal with a large range of activities which are required by the introduction of new products. Some excess capacity is also needed for testing and prototyping-related activities. To achieve fast new product introductions (product response flexibility), the important feature is possibly the integration/coordination between functions involved with the product introduction to ensure design for manufacture (see cases A and B in Appendix 2). This may depend on effective systems and other infrastructural resources - communication channels, organization structure, inter-functions interaction and so on.

Mix flexibility Mix flexibility depends on technological resources which should allow for quick set-ups and should also have a broad capability range. Human resources should also have multi-skills, and infrastructural resources should be effective in order to allow for quick and frequent rescheduling and outsourcing (see case A in Appendix 2). Semi-finished goods stocks can also be used in order to shorten lead-times (assembly to order instead of make to order; see Appendix 2, case D) and allow for quicker mix changes.

Volume flexibility Volume flexibility seems to be related to having excess capacity of human and technological resources or having the ability to get them - through systems - at the amount needed and quickly (see Appendix 2, case D).

Delivery flexibility In order to allow for delivery flexibility it seems that a system should have the same ability of a company with mix flexibility and, additionally, some excess capacity. Otherwise, the setting forward of one order would simply delay others (which is exactly what happens to Company C, which is *bottlenecked*). Delivery flexibility can also be achieved by keeping finished and semi-finished goods inventory (see Appendix 2, case D).

Concluding, apparently no specific type of manufacturing resource plays a major role in developing any of the flexibility types or dimensions. It seems that the type of resource that should be used in order to achieve the type of system flexibility is contingent to the particular situation.

The flexibility of the structural resources

An interesting aspect of the flexibility of manufacturing structural resources was observed in the managers' views, according to which, in terms of structural resources, there is always some sort of resource *reserve* (or *redundancy*) involved in the achievement of manufacturing systems flexibility. Three managers at Company A, for instance, described flexibility explicitly as a reserve, an asset, something which is possessed by the system but which is not being used each time. In fact, if a system is able to respond effectively to a changing circumstance, this implies that the system is able to assume different states and therefore to perform more activities than the activities it is performing each time. It has therefore redundant or excess capabilities. A totally dedicated machine, for instance, is not flexible exactly because it is only able to perform one single task: there is therefore no redundancy in its capability.

Not only redundancy is necessary, though, for the structural resources to contribute to the system's flexibility. They also need to be *switchable* (the term borrowed from Dooner and De Silva, 1990), in order to respond quickly and easily to the changes. In other words, they have to be able to change quickly, easily and cheaply between the activities they are *redundantly* able to perform. See Figure 4.13.

The switchability of the resources may be linked to the system's response flexibility whereas the redundancy may be linked to the system's range flexibility.

There follows an analysis aiming at developing further the concept of resource redundancy and its links with flexibility.

From the field study, it was noticed that some managers consider that structural resources have to possess some level of redundancy or 'reserve' in order to be flexible. To be able to respond to changes in the number of available assembly line workers, caused by absenteeism, Company A, for instance, provides the assembly line with some extra labor capacity. This way, the company covers for the absentees.

Structural resources main contribution to the system's flexibility	Redundancy Switchability

Figure 4.13 Structural resources general contribution to flexibility

This means that Company A's assembly line has redundant *capacity* of the resource labor. However, they also must ensure that the assembly line team has the right skills to perform all tasks, despite the absentees. The way they overcome this problem is by training a number of members of the team to perform a multitude of tasks. As a result of this training, it becomes possible to transfer people between tasks and therefore to accommodate the necessary skills. In providing people with multiple as opposed to dedicated and specialized skills, Company A is creating a reserve, or redundancy of the *capability* of the human resources. Both types of redundancy can also exist in terms of technological resources. A multi-capable machine (such as the ones used by Company A in their CNC cell) has redundant capability, and a production unit with extra machine capacity (such as the one used by Company B in the Zamac injection shop) has redundant capacity.

Besides redundancies with capability and capacity, a third kind of resource redundancy was identified in the field study. Company D, for instance, builds up stocks of semi-finished goods in order to be flexible in responding quickly to its variable demand. To build up these stocks, the structural resources involved are activated before the time in which this activation would be strictly necessary. The build up of stocks allows the system to be more flexible, therefore allowing the company to respond to the changes in demand in a quicker way. It would not be enough for the company to attain this objective only by keeping its current level of extra capacity or extra capability. Aiming at a more rapid response, the system had to activate its resources at an earlier time than would be strictly necessary to respond to firm orders. A stock of parts or products is typically a reserve, built up in order to help the system respond better to a changing circumstance. This reserve is built up by a redundant (or excessive, compared to the needs) utilization of the structural resources.

There would thus be three kinds of resource redundancy, which can translate into resource flexibility, provided that they are managed properly: capability, capacity and utilization. Each of them is further analyzed in turn.

Redundancy in the structural resources capability This is a function of the range of abilities which the resource possesses but which are not being used at each and every time. If a machine, for example, has the capability of performing ten different product or part types, it is more redundant in terms of capability than another one which is able to perform only three different product types (given that both are currently performing one product type at a time). The ability of a machine, expressed as the range of different product types it can perform[14], is in general a design characteristic. Considering the human resources, the redundancy of capability of a worker can be increased by training and/or experience. If a worker is trained to perform a number of different tasks, for example, his capability reserve or redundancy is increased.

Redundancy in the structural resources capacity This is the difference between the level of output the resource is normally producing and the maximum level of output it is able to produce. If a machine has the capacity of manufacturing 1000 parts per hour and is normally assigned to produce 700 parts per hour, it has a larger redundancy in terms of capacity than a similar machine assigned 900 parts per hour. The same concept applies to a worker or to a group of workers.

Redundancy in the resources utilization This occurs when a resource is activated more than was strictly required (such as the build up of stock buffers) or before it was strictly required (such as the build up of time buffers), generating a physical amount of stock. It is basically a redundancy due to the production planning and control system which determined that the excess amount of stock should be produced or purchased at that specific time. Here a stock (generated by a *redundant* utilization of the structural resources) is defined as the amount of raw material, semi finished or finished goods within the system, which has been produced or purchased either in a larger amount or before it was strictly needed to respond to a specific firm customer order.

Manufacturing structural resources redundancy types	Capacity Capability Utilization

Figure 4.14 Structural resource redundancy types

There is another characteristic of the structural resources which is not related to any type of redundancy, but it is also important in the achievement of higher levels of flexibility, mainly response flexibility: the *switchability* of the structural resources.

Resource switchability This relates to how quickly, cheaply and easily a resource switches the activity it is currently performing into another one (Companies A, B, C and D are running programs of set-up time reduction, in order to increase response flexibility, or in other words, technological resources *switchability*). In terms of technological resources, it relates to changeover times which in turn are linked to set-up times. In terms of human resources, it relates to the ease and to the time it takes for the person to switch between tasks up to the point where he or she is performing the subsequent task at the same levels of performance as he or she was performing the previous one.

Summarizing, a structural resource is flexible as long as it has the appropriate amount and types of redundancy and levels of switchability which are required in order to respond effectively to the system's needs.

Structural resources main contribution to	System Flexibility	Redundancy	Capability Capacity Utilization
		Switchability	

Figure 4.15 Structural resources contribution to system flexibility

The flexibility of the infrastructural resources

The system's flexibility is not a function exclusively of the redundancy and switchability of its structural resources. They are necessary but not sufficient for a system to achieve flexibility.

In terms of infrastructural resources, it does not seem necessary for the system to have any redundancy or excess, in order to achieve higher levels of flexibility. Infrastructural resources have to play their role *precisely* to contribute to flexibility: establishing effective inter-function communication and a participative and agile decision-making process and ensuring that the appropriate information, at the right level of detail, gets to all the decision points which need it, accurately and as soon as possible. An example is the simultaneous engineering which helps in the achievement of new product response flexibility (see Appendix 2, case A). Two systems with the same level of redundant resources (e.g. spare capacity for prototyping) will produce different performance in terms of reacting quickly to a customer's request for a new product if they have different levels of ability to make the right information get to the right decision points the earliest possible. If one coordinates well the product design function with the process design function, for example, and the process design function learns of the relevant information they need at early stages, the process design function will be able to start their work in parallel with the product design work, and the whole system as well as the customer are likely to benefit from the parallel - as opposed to sequential - activity development, in terms of response time.[15]

Infrastructural resources do not seem to influence range flexibility as dramatically as they influence response flexibility. Infrastructural resources only have a supportive and facilitating role in the manufacturing process. The structural resources are the ones which actually 'do the job' and therefore they are the determinant resources in terms of the range of possible jobs or activities the system can perform.

The control-flexibility relationship: a systems approach

The set of resources of technology, labor and infrastructure in a manufacturing plant work as a system. The manufacturing system performance is a function of its specific configuration of resources although it seems plausible that different configurations can achieve similar performance levels. Determined levels of flexibility of the manufacturing system, for instance, could be achieved by different configurations of particular redundant and switchable structural resources and different types of infrastructural resources.

When an organization decides what type and amount of control it is going to exercise over its stimuli, it is also and automatically defining the stimuli which will be 'allowed to pass the control filter'. In other words, the organization is also defining what sort of changes it is going to deal with or respond to, either because the ability to deal with some changes is strategically important (flexibility may be important as one of the system's competitive criteria) or because it is economically inconvenient or even impossible to control them. By being flexible, companies handle the effects of the 'non-controlled' stimuli. A typical example is the occurrence of unplanned changes with the machine availability caused by breakdowns. All the case companies' managers emphasized that such changes should be avoided (controlled) by developing preventive maintenance procedures. However, since it is impossible or sometimes inviable to eliminate completely the occurrence of machine breakdowns via prevention, it is necessary that the system is flexible, or in other words, it is able to respond quickly once a breakdown occurs - or once a stimuli 'passed through the control filter'.

Control here is defined as every activity which aims at interfering with the amount and/or nature of the stimuli with which the system has to administer. Control activities are developed before the occurrence of the stimuli. Once the stimuli happens, there is nothing that control activities can do to manage it. This is the point where flexibility comes into the scene in order to respond to the change, utilizing the redundant and switchable resources the manufacturing system possesses, as quickly as the infrastructural resources are able to make the relevant information about the stimuli reach all the appropriate decision points.

Sometimes, the same action, say coordination, can serve both purposes - to increase the unplanned change control and to increase the flexibility. Consider coordination with suppliers. On the one hand, coordination can have a character of control, reducing the uncertainty the organization works under, with regard to the suppliers. On the other hand, coordination with suppliers can also have the purpose of increasing the organization's flexibility, making sure that the relevant information about the changes in its demand pattern, for instance, are acknowledged rapidly by its suppliers, and therefore the response of the *system* formed by the organization and its supplier can be quicker. This is an interesting aspect of the relationship between control and flexibility, which is related to where one defines the system's borders. Control (such as coordination between two departments) at one level (internal) can represent flexibility from the viewpoint of the external environment or of the next external level (for instance of the system's customer). When the company's product demand changes, for example, the response to that change will probably be faster and better (or more flexible) if there is a good level of coordination between the company and its suppliers. This is 'intra system' coordination which does not interfere with the change in demand itself. It aims at increasing the overall system's *flexibility*. If the company, on the other hand, develops coordination or negotiation with its customer aiming at working with more stable schedules, that is *inter system coordination* aiming at increasing the company's *control* over the demand changes in terms of reducing its frequency (by freezing schedules, for example) or its predictability (by having a longer planning horizon, for example).

The double character shown in the example above does not seem to happen exclusively with coordination. By improving forecasting systems, for instance, a company can reduce the level of uncertainty it works under but at the same time it can prepare itself better to respond quicker to future customers orders (such as Company B's build up of 'strategic finished goods inventories') and possibly increase the flexibility which the customer perceives. This finding seems to be in accordance with the idea of the need to ensure some baseline stability (or 'rigidity') in order to allow for flexibility, advocated by Adler (1987) who argues that the point about flexibility is not to increase flexibility indefinitely, but to find the right mix of rigidities and flexibilities.

The examples above suggest that in order to analyze flexibility and control we should adopt a systems approach. Intra system control, for instance, can make for inter system flexibility. Intra system control aims at reducing the intra system turbulence caused by the stimuli. The same managerial action can have a character of control or a character of flexibility depending on where one defines the system's borders. These considerations are only possible if a system's approach is adopted.

Summary of the main aspects of the proposed model

There are two main types of change affecting the manufacturing systems: planned change and unplanned change. This model is primarily concerned with the management of unplanned change, which is called stimuli here. Stimuli or unplanned change has five main dimensions: size, novelty, frequency, certainty and rate.

Managers use two main approaches in order to deal with unplanned change: either they try to *control* the amount of unplanned changes which affect the manufacturing system operation by acting before the occurrence of the change or they try to be *flexible* by developing the system's ability to respond effectively to the unplanned change after its occurrence.

Seven general types of managerial actions which represent ways of exercising unplanned change control were identified: monitoring and forecasting; coordinating and integrating; focusing and confining; delegating and subcontracting; hedging and substituting; negotiating, advertising and promoting; and maintaining, updating and training.

Five general types of manufacturing systems flexibility are important for a response to the unplanned changes which were left uncontrolled because it was either impossible or inconvenient to control them: new product, mix, volume, delivery and system robustness flexibility.

Infrastructural resources seem to play a major role in terms of exercising unplanned change control whereas no resource type was considered to be particularly more important than the others in terms of developing flexibility. It is suggested that the answer to the question 'what type of resource should be developed in order to achieve what type of system flexibility?' depends on each particular contingency.

Manufacturing systems flexibility is developed through the process of a configuration of flexible resources. For there to be flexibility, structural resources have to possess some level of redundancy in terms of their capacity, capability and/or utilization and some sort of switchability, which is the ability to switch easily, cheaply and quickly between tasks. Infrastructural resources do not have to be redundant in order to contribute to the system flexibility, they only need to perform their function properly.

It is essential to adopt a systems approach in order to understand properly the concepts in isolation and the complementarity between *change control* and *flexibility*. Intra system change control, for instance, can contribute to the flexibility of the overall system.

Notes

1 According to Eisenhardt (1988), contrary to popular thinking, one of the key features in theory building research is the initial definition of the research problem, at least in broad terms. Although no existing theories are in consideration in the present research and no formal hypotheses are being tested, some a priori variables are considered which are likely to be relevant in the theory building exercise. Miles (1979) considers that research projects that intend to come to the study with no assumptions usually find much difficulty. The author believes that at least a rough working frame needs to be in place at the beginning of the fieldwork.

2 The level of analysis is the level of production units (see Appendix 1 for details).

3 A computer software-based manufacturing planning and control system.

4 This is the definition given in *The Oxford Paperback Dictionary*, third edition. Oxford University Press. Oxford, 1988.

5 See Chapter 1 for a discussion on manufacturing focus.

6 These are groups of machines, generally in charge of completing one or several families of parts. See Burbidge (1989) for details.

7 Slack (1987) observed that managers would try to avoid being flexible, in his empirical work. He however does not explore this idea further.

8 These are production control systems in which downstream operations consumption of materials triggers upstream operations production, 'pulling' material throughout the production process.

9 See Schonberger (1982) for a detailed description of the Kanban technique.

10 See Appendix 2, case C, for details.

11 Focused manufacturing relates to focusing the operation on a limited task by selecting a limited, concise, manageable set of products, technology, volumes and markets to be served while structuring basic manufacturing policies and supporting services so that they focus on one explicit manufacturing task instead of many inconsistent, conflicting, implicit

tasks (Skinner, 1974). See chapter 1 for a discussion on manufacturing focus.

12 This is the capacity to continue functioning effectively despite the change (Mandelbaum, 1978).

13 See Goldratt (1990), for a detailed discussions on bottleneck buffering.

14 This is just a simple example. Other considerations are also important in assessing the capability of a machine, such as how different are the products it is able to produce.

15 See Womack et. al. (1990) for a discussion on the simultaneous development process.

5 Conclusion

Objective

The objective of Chapter 5 is to summarize the main findings of this research work, compare them with the current literature and suggest research avenues and opportunities which still lack further exploration. The first part of the Chapter concentrates on the empirical findings, drawn from the field study which was realized as part of this research and which is described in Appendix 2. The second part concentrates on the principal aspects of the conceptual model proposed. The model development process is described in Chapter 4. The third part of Chapter 5 suggests some research opportunities which are open or which have not been explored sufficiently by this research.

The main empirical findings and the current theory

In this section, the main empirical findings of this research (see Appendix 2 for details) are presented and briefly discussed, and some comments and comparisons between them and the current literature are also presented.

Managers consider flexibility as one of the ways to deal with change in organizations, mainly when the change perceived by the organization cannot be controlled (eliminated or reduced).

Some authors (e.g. Mandelbaum, 1978) have proposed that flexibility can be used in order to deal with changing circumstances. However, little empirical evidence has been found in the literature to back this proposal. Slack's (1987) work represents an exception and resulted in his hierarchical model of manufacturing flexibility. However, the main focus of Slack's research was on flexibility and therefore the control of the uncertain and variable changes which the manufacturing systems are subject to, or in other words, the managerial alternative of trying to reduce or eliminate them, was not contemplated. The present research findings can be regarded as an extension of Slack's model. They constitute an attempt to include not only the managerial actions which aim at dealing with the effects of unplanned changes after the change, but also to consider interference with the amount of unplanned change which the organization has to handle before the occurrence of the change. No similar approach has been found in the literature, although the proposed approach is not contradictory with most of the existing research work in the fields of flexibility and/or control. Rather it is an attempt to organize and define more precisely both concepts; the aim is to help managers understand and analyze the manufacturing system's functioning from an alternative viewpoint.

Managers do not always discriminate explicitly between control and flexibility and do not always have a clear view of what should be the most appropriate way to deal with the different types and dimensions of unplanned change.

The approach generally adopted by the literature has not helped the managers much in making the formal distinction between control and flexibility. Invariably, both concepts are treated in isolation and under a variety of labels. The literature on 'management of change' contemplates primarily the management of planned change, neglecting the type of change on which we focus in this research: the management of the unplanned changes (called here *stimuli*). No relevant research work has been found on the classification of unplanned change as such. The literature on 'uncertainty', by some theorists on 'organizational behavior' attempted to classify uncertain changes in the 1970s but the research on the field seems to have come to a halt after some public disagreements between some researchers via less than complimentary replicas and treplicas to each other's articles (see Chapter 3). Apart from that, the 'non-uncertain' changes are not contemplated in their analyses either. The mention made by some interviewed managers regarding the need to deal with change as such, sometimes regardless of its certainty, led us to rethink the tentative 'motif' which we had initially considered, based on the literature, for a system to be flexible: from the consideration of uncertainty and/or variability to the consideration of change and its dimensions.

Managers understand that different types of change should be dealt with by developing different types and characteristics of resources. However, in general they do not appear to have a consistent model to help them make decisions in that regard.

That is consistent with the approaches found in the literature, which are generally contingential. Several authors attempted to suggest types of flexibility which would be appropriate for dealing with different uncertain changes or with the variety of the system's outputs. However, possibly because of the use of an inappropriate flexibility taxonomy or because of lack of empirical research to back them, the models in the literature which attempted to relate types of, say, uncertainty and types of flexibility (for example, Gerwin's, 1986) are very general and of little practical use. They associate, for instance, 'uncertainty with overall volume' with 'volume flexibility' (see Chapter 2 for more examples), failing to analyze which managerial alternative actions are involved in developing, for instance, volume flexibility. In this regard, in order to deal with the uncertainties and with the variability of outputs they have to face in their day-to-day activities, the managers identified several methods during the interviews. A more detailed discussion on them and on the related literature follows.

The literature suggests links between uncertainties and flexibility (Gerwin, 1986; Slack, 1990a), but it lacks empirical evidence on the issue. The present research work provides some empirical evidence that flexibility is actually one of the ways in which managers deal with uncertainties.

Gerwin and Tarondeau (1989) propose the following relationship between uncertainty types and flexibility types:

Uncertainty with regard to:		
Desegreggated product demand	needs	mix flexibility
Product life cycle	needs	changeover flexibility
Product specification	needs	modification flexibility
Aggregate production	needs	volume flexibility
Machine downtime	needs	routing flexibility
Process characteristics	needs	specification flexibility

On top of the fact that some relationships are merely trivial, as is the case with 'volume flexibility is needed to deal with demand volume uncertainties', the authors' prescription seems to be of limited use for managers. The authors do not analyze the relationships further, in terms of how the managers should go about reaching, for instance, modification flexibility (or the ability of a process to make functional changes in the product, according to Gerwin, 1986) or mix flexibility (the ability of a manufacturing process to produce a number of different products at a certain point in time: Gerwin, 1986), which is not trivial.

Consider the last case, as an illustration. Suppose there is a manufacturing system with a very large number of small and totally dedicated machines, with very long set-up times. This hypothetical system is surely able to produce 'a number of different products at a certain point in time'. But is it flexible in mix? Can it really deal with uncertainty with 'disaggregated product demand'? It is doubtful that it can. The authors failed to recognize and prescribe that, in dealing with uncertainty in mix, the most important characteristic is exactly the switchability of the resources, provided that there is sufficient capability in the process. That is an example of the reasoning that led us to try to link types of uncertainty to resource characteristics (see Table 5.1) rather than to systems flexibility types. The managers with whom we talked also seemed to be more used to thinking in terms of managerial actions or features referring to resources, as opposed to categories or classifications related to the whole system functioning which, most of the time, they were not aware of . (An example of this is mix flexibility.)

Apart from that, some of these actions or features can, according to the managers, have an effect on a number of system flexibility types. (For example, a multi- capable machine can help in Gerwin's mix flexibility and also in routing flexibility.) By the same token, in order to achieve one of the types of system flexibility, it is frequently necessary that more than one feature concurs. (For instance, for routing flexibility to be achieved, multi capable machines alone are not sufficient; it is also necessary that the system 'knows' or has information such as the alternative routes of the products or parts involved.) This inter relationship between resource features, managerial actions and system flexibility types makes it dubious to talk about what system flexibility types should be used in order to deal with which different types of uncertainties. The explanation for this statement follows in the form of a hypothetical situation. Consider that a specific resource feature RF is necessary in order to deal with a certain type of uncertainty U. Suppose that this resource feature RF helps (but does not determine) some system flexibility type FT. To assume that being flexible in that system flexibility type FT is what is required in order to deal with the uncertainty U is, to say the least, risky. The reason for this is that being flexible in that system flexibility type FT does not necessarily mean that the specific needed resource feature RF is possessed by the system, since RF is not the only determinant of the system flexibility type FT and it could even not be present in the system if other alternative determinants are.

Knowing the specific resource features which are desirable in order to deal with the different uncertainty types seems therefore to be a more appropriate approach. One can always relate the resource features to the system flexibility types *a posteriori* if necessary or convenient. The following are the ways that managers of the case companies mentioned as appropriate for dealing with the various types of uncertainty they had previously mentioned as important:

Table 5.1 Ways of managing uncertainties

Uncertainties regarding...	can be dealt with by developing...
parts and materials supply	rescheduling ability coordination with the suppliers buffer stocks internal machine capability
product mix demand	ability to re-schedule production fast set-ups stocks of finished and semi-finished goods ability to get short lead times with suppliers
machine breakdown	preventive maintenance fast corrective action re-routing capability
labor absenteeism	labor multi skills some excess capacity of labor
new product introduction	integration design/development/production ability to subcontract supply
management behavior under change	training / awareness improvement
global demand	forecasting systems
labor supply	internal training
government intervention	shorter cycle times
technology information	ability to subcontract supply
unions behavior	close environment monitoring

Although the managers were asked about how they dealt with the uncertainties after the uncertain changes had already occurred, many of them emphasized measures which should be used in order to avoid the occurrence of the changes. In the Table 5.1, many relationships actually refer to control (e.g. preventive maintenance and development of forecasting systems) rather than flexibility. This behavior of the managers is possibly related to the new ideas brought about by concepts such as 'lean production', which advocates proaction rather than reaction for the manufacturing function's attitude. Womack et al. (1990) illustrate this point when describing aspects of the organization of the 'lean-factory':

> So, in the end, it is the dynamic work team that emerges as the heart of the lean factory [...].Workers then need to acquire many additional skills: simple machine repair, quality checking, housekeeping and materials-ordering. Then they need encouragement to think actively, indeed proactively, so they can devise solutions before the problems become serious. *(Womack et al., 1990)*

The managers who are more aware of manufacturing flexibility see flexibility as a 'reserve', something which should be planned for, developed, maintained and considered as a valuable asset.

An approach to flexibility put this way has not been found in the literature. Slack (1989) is one author who actually suggested the existence of some sort of redundancy in flexible resources but he did not attempt to analyze it further, in terms of establishing, for instance, types of redundancy. One of the advantages of modeling flexibility as a reserve is that the idea, when presented *a posteriori* to some production managers in courses on manufacturing flexibility, given at the University of Warwick and other institutions, appeared to be very appealing to them. The proposed model was informally presented to manufacturing managers from a number of companies (belonging predominantly to the automotive and aerospace industry) and they were then asked to comment on it. Generally the comments were favorable, in terms of modeling flexibility conceptually.

Flexibility, when presented as such during the interviews, with its character of 'potential', had sometimes seemed to be too abstract a concept for the most pragmatic managers. Possibly they feel more comfortable with a concept such as a reserve because they are in a way used to dealing with 'reserves', e.g. inventory (reserves of material) and money (reserves of capital).

The proposed model and the current theory

The main points of the proposed model are discussed below (see Chapter 4 for details of the proposed model).

In order to manage manufacturing systems effectively, it is important to understand better the concept of change in an increasingly turbulent competitive environment. The current literature approaches change invariably in a partial manner. Either it concentrates on planned change (such as the organization behavior and development approach) or, it treats unplanned changes from an invariably partial and segmented perspective.

Two large streams of research can be identified on the issue of managing unplanned change. In general, one stream is found under the label 'flexibility' and, although sometimes not rigorously defined, it aims at dealing with the change and its effects after the fact. In this stream, the literature's approach is also generally partial (with few exceptions), concentrating on specific resource types (such as machine flexibility or labor flexibility) and failing to analyze flexibility at the manufacturing system's level consistently.

The second stream, indirectly, aims at reducing the amount or nature of the changes which the system must handle. Several management techniques and research fields are engaged in finding ways of controlling the dynamics and the magnitude of the changes which affect the manufacturing systems; forecasting techniques, maintenance systems, parts standardization techniques and manufacture focusing are some among numerous examples. Their aim is to attempt to avoid the occurrence of the change, before the fact or in other words, preventively.

Although the purpose of both streams is to manage unplanned change, the current literature lacks a unifying framework which encompasses both streams and helps managers understand and analyze the concepts of unplanned change, control and flexibility and their inter relation. The present research proposes such a framework. Its main propositions and their relation to the current literature are summarized below.

The main propositions of the proposed model are:

Stimuli, or relevant unplanned changes, to which the manufacturing system is exposed, have five dimensions: size, frequency, novelty, certainty and rate. It is important to classify stimuli because different stimuli dimensions may call for different managerial actions.

The current literature is very fertile when dealing with planned change. Numerous publications can be found on issues relating to 'how to change the organization effectively', under various labels; organization development, organizational behavior and management of change are some examples. However, the literature is scarce in terms of unplanned change. Flexibility is possibly the only research field where dealing with changing circumstances is an explicit objective. However, interestingly enough, despite mentioning change when defining the field, the authors do not appear to dedicate much attention to understanding and classifying change when developing the theme. Instead, some of the authors concentrate on some aspects of the changes such as uncertainty and variability of outputs.

Both variables are included in the framework proposed here, since one of its proposed change dimensions is exactly the uncertainty of the change. Variability in the framework is translated into the change dimensions of novelty, size, frequency and rate.

There are two basic and complementary ways of managing stimuli in manufacturing systems: by controlling the stimuli and by being flexible, or by dealing with the stimuli's effects.

The following definitions are proposed:

Control is defined here as the ability to interfere effectively with the causes of the changes or with the way the system senses the changes, in order to alter one or several of the dimensions whose effects the system will otherwise have to respond to.

Flexibility is defined as the ability to deal effectively with the effects of the unplanned changes, as these effects are experienced by the system.

This complementarity between control and flexibility was suggested by Gerwin and Tarondeau (1989) in terms of dealing with uncertainties, but it was not formalized. It is also being proposed here that this approach can be extended to deal with all dimensions of change and not only with uncertainty.

The proposed approach also allows for a somewhat more rigorous definition of manufacturing flexibility. Some loose definitions of flexibility can be found in the literature. Buzacott and Mandelbaum (1985), mentioned in Gupta and Goyal (1989), define flexibility as 'the ability of a manufacturing system to cope with changing circumstances'.

Strictly speaking, if control is not considered, flexibility should not be regarded simply as the ability to cope with changing circumstances, but rather, to respond to changing circumstances. The reason is simple: in a hypothetical situation, avoiding involvement with a change in demand for instance, can be a good way to cope with the change. It can possibly even be good managerial practice, but it is hard to consider such practice as flexibility. That would be, for instance, the case of the corporate strategy of a hypothetical organization which focuses on a stable standard line of products aiming at a specific niche of the market, in order to prevent the manufacturing system from being exposed to frequent and novel demand changes. The adaptation may have been effective, but that does not make the manufacturing system flexible. One dictionary definition seems to catch the point very well: 'flexibility is the ability to be changed to suit circumstances'. (*Oxford Paperback Dictionary*, third edition). The idea of being changed implies that the system is actually reacting and not avoiding having to react.

In order to conceptualize flexibility rigorously, it seems to be important to bear in mind the complementarity of its concept with that of unplanned change control, which is scarcely addressed in this way in the operations management literature.

The unplanned change control methods used by the system thus would work as a filter, restricting the amount of change effects the system has to deal with. The changes which 'pass through the control filter', then, must actually be dealt with by the system, through its system flexibility characteristics.

Intra system unplanned change control can be used in order to improve the level of the system flexibility.

Because unplanned change is in general disrupting and the level of flexibility of a system is also dependent on the speed with which it reacts to changes, some intra system change control can help increase the system's flexibility, mainly in terms of its responsiveness. Many examples of intra system change control can be found in the literature. Manufacturing focusing, preventive maintenance and manufacturing planning and control systems are all systems which attempt to control one or several dimensions of the changes within the system. This intra system control can possibly help increase the levels of flexibility of the system as a whole (or the inter system flexibility), because it allows for the system to be more responsive (e.g. via coordination) and less exposed to undesirable disrupting set backs.

This makes the flexibility-control relationship more complex, and the systems approach-dependent and it demonstrates once more the importance for the manufacturing systems managers to understand it well. The literature has not addressed this relationship explicitly.

There are five types of flexibility which are relevant for the analysis of manufacturing system's flexibility: new product, mix of products, volume, delivery and system robustness flexibility.

In general, the literature on flexibility emphasizes the flexibility of particular resource types. Although these approaches can be useful to help address specific problems such as the comparison between flexibilities of two machines, from an operations viewpoint they are of limited use, because they usually do not link the flexibility of the resources with the strategic objectives of the manufacturing operation. To be able to do this, a systems approach for analyzing flexibility appears to be more adequate, in which all the resources are analyzed and considered as the interacting elements which form the manufacturing system. Some research found in the literature addresses flexibility at this level (which allows for analyses of the contribution that a flexible system can give to organization competitiveness), but it concentrates on the application of flexibility concepts with the intention of allowing for the system to be able to change its outputs. However, there is no treatment of the use of flexibility concepts for preventing the system from being disrupted from unexpected setbacks with the supply side and with the process itself. These authors leave it somewhat implicit that if a system is flexible in terms of changing its outputs it would automatically have developed abilities to overcome setbacks with the process and inputs. We demonstrate that this is not always valid and that in some situations specific flexibility-related abilities should be developed in order to achieve desired levels of system robustness.

Some authors, on the other hand emphasize the importance of flexibility in ensuring that the system's functioning is 'protected' from the occurrence of these setbacks but they have not developed the concept further in terms of its application. Mandelbaum's state flexibility and Buzzacott's machine flexibility, for instance, both relate to the ability of the system to cope with changes and disturbances for its constituent elements. However, in their research work, the objective was to conceptualize types and dimensions of flexibility rather than to analyze the operationalization of such flexibilities. They also lack empirical evidence.

This research proposes a fifth type of system flexibility to be added to the four proposed by Slack (1989), types considered appropriate for the analysis of demand-related flexibilities. The fifth type is called here 'system robustness flexibility'. The proposed type has particular characteristics and aims at filling the gap identified above. The ways some managers operationalize such a concept are also discussed in Chapter 4, drawing from information from the field work.

There are several ways of exercising control over the unplanned changes. Some of them, drawn from the literature and from the field work findings, are represented by the following general managerial actions: monitoring and forecasting; coordinating and integrating; focusing and confining; delegating and subcontracting; hedging and substituting; negotiating, advertising and promoting; and maintaining, updating and training.

The literature is fertile in ways of exercising control over the unplanned changes. Several managerial techniques aim, although not explicitly, at reducing one or several of the aspects or dimensions of the changes the system would otherwise have to deal with.

Some of them can be identified:

Vertical integration This is a way of achieving control through *coordination and integration*. The literature on vertical integration is vast. One of the reasons for a company to vertically integrate, according to the literature, is to reduce transaction costs. Putting it simply, transaction costs would happen owing to reasons such as imperfections in the communication between the company and its suppliers and the less than complete and precise information about the supplier's processes. By integrating vertically, the company would improve communication and would also have more information about the supplier's behavior and process. Vertical integration is therefore a way of exercising control over changes in the relationship with the suppliers.

Forecasting techniques These are techniques which are used for achieving control through *monitoring and forecasting*. Quantitative (regression analysis, time series, among others) and qualitative methods (e.g. some market research methods) are numerous in the literature and they are used in order to try to predict and anticipate future events in an attempt to reduce the uncertainty and novelty aspects of the future changes.

Manufacturing planning and control systems These are attempts to increase control through *coordination and integration*, trying to coordinate demand with supply from the customer's demand along the internal supply chain and sometimes extending the coordination to external suppliers. MRP II system and Kanban system, for instance, are different systems aiming at least one common objective: controlling change along the organization's supply chain.

Preventive maintenance One more particular case of one of the proposed types of control - *maintaining, updating and training* - preventive maintenance aims at 'building reliability' on the equipment, aiming at the reduction of the uncertainty, frequency and/or impact (size) of the changes brought up by machine breakdowns.

Suppliers development A particular case of *maintaining, updating and training*, supplier development aims at building reliability on the supply base, in order to control the frequency and uncertainty of problems regarding supplies, in terms of delivery dates and quality levels.

Make-or-buy analysis This is related to *delegating and subcontracting*. Companies can control change dimensions by deciding, for instance, to buy in a component of which the technology is changing extremely fast. In this case, the uncertainty and rate of the possible changes regarding the involved technology would be controlled.

Standardization From an operations viewpoint, standardization is a particular case of *negotiating, advertising and promoting*. Standardization must generally be negotiated with the manufacturing internal customers (such as the marketing function) and suppliers (such as the design function), and intends typically to reduce the frequency of the changes (such as the machine changeovers) with which the manufacturing function has to deal.

Simultaneous engineering It is a particular case of *coordination and integration* between functions of the organization. People responsible for the several stages of development of a new product get together at early stages and coordinate their efforts to make the stages as simultaneous as possible, as opposed as the traditional sequential fashion. People from different functions such as production and process and product development, cooperating at early stages, can also ensure that things are designed and made right the first time, reducing the frequency of future changes.

Manufacturing focus It is a particular case of *focusing and confining*. By focusing production units on specific manufacturing tasks, the organization can either control the frequency of the changes (if it chooses to focus on large orders), the novelty and certainty of the changes (if it chooses to focus on standard parts) or the size of the changes (if it, for instance, chooses to focus on a specific range of order sizes).

The infrastructural resources play a major role in exercising control over changes.

This is demonstrated by analyzing the control types (see Chapter 4 for details). All of them relate in one way or another to systems (e.g. information systems, organizational systems, control systems and supply systems).

The contribution that the infrastructural resources can give to the manufacturing system flexibility is related to response rather than range flexibility, which is primarily dependent on the structural resources.

The literature on flexibility also does not conceptually discriminate the roles that structural and infrastructural resources play in the achievement of system flexibility. Such discrimination is important because in order to achieve desired levels of specific flexibility types, specific resources have to be developed in specific ways.

The contribution that structural resources can give to the manufacturing system's flexibility is through structural resource redundancy - in terms of capability, capacity and utilization, and switchability.

The idea, drawn from the field study, that flexibility can be seen as an asset of the manufacturing system, as well as Slack's (1991) suggestion that flexibility implies some sort of redundancy of resources, motivated the development of the 'redundancy model'. The analysis of the redundancies of the different resource types lead to the understanding of the differences in the role of structural and infrastructural resources in terms of supporting the system's flexibility. Because the redundancy model is original, the literature does not comment on it.

As a by-product of the redundancy model, the role of the stocks in terms of their relationships with systems flexibility has been clarified. According to the approach in the present research, there is not a clear conceptual difference between holding excess capacity, having abilities in excess and holding stocks of goods. All are manifestations of resource redundancies, in terms of capacity, capability and utilization, respectively. The three of them are elements with which the managers can play in order to achieve manufacturing system's flexibility.

The current literature lacks a model which explains the relationship between flexibility and stocks, frequently seen as mutually exclusive alternatives. The complementary rather than exclusive alternatives approach proposed here does not prescribe or recommend the use of stocks of goods in order to achieve flexibility, but it is an attempt to call the attention to the fact that stocks are a possible complementary alternative to other ways of achieving flexibility. There are trade-offs to be considered in the process of configuring the resources, which should take into account all the costs involved with holding stocks, holding capacity in excess and holding capability in excess.

A look forward: some unanswered questions

There are some questions regarding the relationship between uncertainty, variability of outputs and flexibility in manufacturing systems which are still to be answered. Some of these questions are discussed below.

The costs of flexibility and control and the trade-offs involved

The amount of control and flexibility used by a manufacturing system in order to manage its unplanned changes is, to a certain extent, a managerial choice. Although they are not the only considerations to be made, there are trade-offs to be considered between control and flexibility. In order to analyze such trade-offs, it is important that the managers have a good understanding of the costs of controlling the unplanned changes and the costs of developing flexibility and also of being flexible. This is an issue which is scarcely explored by the literature and surely needs further development.

The relationship between system flexibility and resource flexibility

The relationship between desired or required system flexibility levels, set by the manufacturing strategy, and the system's resource characteristics which are necessary in order to achieve them is something which, although addressed briefly in this research, needs further exploration. The literature, in general, does not discriminate properly between different levels of analysis with regard to flexibility. In some of the authors' lists of flexibility types, one can find flexibility types from two or even three levels - the machine, the system and the organization as a whole.

It is important to have a consistent set of flexibility types and dimensions, which can be linked to the organization's strategic objectives, when analyzing the flexibility of manufacturing systems. The model proposed here provides this classification. However, given the levels of flexibility which the particular manufacturing system has to achieve, which resources should be developed in order to achieve such levels? This is a question which has not been sufficiently explored either in the literature or in the present research work and is certainly an issue which deserves further attention.

The costs of the different redundant resource types

For the appropriate mix of required flexibilities to be achieved, choices of the adequate configuration of resource redundancy should be made. Some choices are quite clear. To achieve product range flexibility, a firm has to use its redundant capability because neither goods nor capacity will help. In some situations, though, managers do have alternatives to choose between. For example, if a system is being designed to have a highly flexible response to volume changes, some alternatives are available: redundant stocks may be used as well as redundant capacity or still a mix of both. If a system needs high mix response flexibility, a choice must be made between very flexible machines and workers and some level of stocks of finished and semi finished goods. The trade-offs involved must be considered for each and every situation contingently. At the system's level, therefore, a plant can be flexible via different configurations of the three types of redundant resources. Alternatives at the resource level represent trade-offs to be done at the system level. The present research has not explored this aspect and the issue should receive further attention.

Contingency relationships within the model

The contingent aspect of the proposed model appears to be clear. There are specific actions to be taken, decisions to be made and resources to be developed which are contingently more appropriate for dealing with specific environmental and internal conditions and objectives. However, the contingency relationships between the categories within the model still require further development:

i) between the unplanned changes dimensions and the control types and flexibility types and dimensions;

ii) between the types of stimuli control and the specific structural and infrastructural resources; and

iii) between the types of system flexibility and the types of structural resource redundancy and switchability and infrastructural resource types.

An audit procedure based on the proposed model

The development of an audit procedure based on the proposed model seems to be important for the managers to be able to make further practical use of it. The procedure should include:

i) the identification of the key types of change to which the system is subject, be they internal or external, related to the inputs, process or outputs. There should be consideration of both types, those changes which represent opportunities and those which represent risks for the company's present and future competitive position, considering the manufacturing strategy of the company; and

ii) the evaluation of the flexibility-related performance of the manufacturing system and resources. The model developed in this research can help managers decide how they should cope with the identified changes to which the company is subject. This way, comparisons between what is being done and what should be done could be made and actions could be planned based on the discrepancies.

A rationale behind the unplanned change control types

The types of change control proposed deserve further attention. The list of seven types proposed does not intend to be exhaustive. More types may be found and, in addition, it seems possible to determine a general rationale behind the seven types.

A look back: a critical review of this research work

Some points are worth mentioning in an attempt to evaluate with hindsight the way in which this research work has been conducted. Such an evaluation can be of some help for researchers and students in order that they can avoid repeating mistakes and that they can take advantage of the parts which proved to be successful.

The case-study approach

The case study with the semi-structured interviews approach used as the research method proved to be adequate for the objectives of this research. It allowed the researcher to adjust and to a certain extent, redirect the focus of the research, from the one defined at the outset, which aimed at exploring the relationships between flexibility, uncertainty and variability to the broader view of managing unplanned change. It was then possible to build theory through the development of a model which is described in Chapter 4. The semi-structured interviews allowed the researcher to explore aspects which were not present in the research propositions defined at the outset of the project, such as the concept of stimuli control and its types, which ended up being one of the central aspects of the proposed model. The same is true with regard to the treatment of flexibility of the manufacturing resources using the idea of resource redundancy. This idea would probably not have been developed if a survey had been conducted based, for instance, on a structured questionnaire sent to the managers by mail, or in other words, without the presence of the researcher. The case study approach also allowed for the choice of polar and rich cases, because the cases were not chosen at random (see Appendix 1). The companies in Brazil, as expected, provided polar examples which allowed the exploration of the concept of environmental uncertainty and ways the managers deal with it and the British companies provided examples of rich product variety and modern approaches to it. A random sample, which would probably be the case if the chosen approach had been a large survey, would possibly not contain such rich and polar cases.

The difficulty in obtaining hard data about flexibility

Although already considered an important feature of manufacturing systems by most of the managers interviewed, flexibility is still a concept which is rarely measured and accompanied in the case companies. Most of the managers agreed that they should attempt to measure the system's performance in terms of flexibility but at the time of the research field work none of them had implemented procedures which measured important aspects of the ability of the manufacturing system to change. Hence the scarce hard data present in this research work which relied heavily on the manager's perceptions. Hopefully, in the near future the organizations will develop procedures in order to assess and accompany the flexibility of their manufacturing systems and so more objective analyses will be possible for future research.

The question of results generalization

There is no evidence that the results achieved are not analytically generalizable beyond the limits of the case companies analyzed and even beyond the automotive industry, to which the case companies belong. The present research did not aim at statistical generalizations about a population, but at analytical generalization. The categories present in the proposed model seem to be sufficiently general to be also suitable for other industries and manufacturing types. The case companies were chosen within the metal engineering automotive industry because they represent possibly one of the most complex industries in terms of operations management. Other industries will probably have less complex problems in managing their unplanned changes, but hopefully the richness and the complexity of the cases analyzed in this research will also help their managers to understand better the concepts involved.

A summary of the main apparent problems with the methodology used

The sample is possibly not representative of the case companies industry and the findings are not statistically generalizable In order that the results could be considered statistically generalizable, the sample should have been possibly much larger and chosen at random. In this case, however, the richness of the polar cases could have been lost. The research did not aim at statistical generalizations from the outset. The main aim, which has been achieved, was to build theory which should be analytically rather than statistically generalizable.

The findings are largely based on the managers' perceptions which can be biased Given the objectives of this research, however, any other method would probably have even more problems, because the variables involved in this research, such as uncertainty and flexibility, are complex and the methods for their objective measurement found so far in the literature are controversial and highly arguable (see Chapters 2, 3 and Appendix 1 for discussions on the measurement of flexibility and uncertainty).

The treatment of the data included the interpretation, by the researcher, of the opinions expressed by the managers, which can also bias the findings Nevertheless, given the need to use the manager's perceptions, no alternative was left except to interpret the manager's answers. If a closed structured questionnaire had been used, in order to avoid interpretation of the answers, the exercise of building theory would be jeopardized, since the managers would not have the opportunity to extend their comments on the researched topics. Developments such as the 'redundancy model' of flexibility (see Chapter 4), among others, would possible not have been achieved. Additionally, the data treatment process was as systematic and careful as possible in order to try as much as possible to avoid biasing the findings.

The choice of the case companies and the managers to be interviewed was done arbitrarily and therefore there is no guarantee that it was the best choice possible Once again, the alternative would have been a random sample, which, given the resource constraints of the researcher, could not have been large, given that the presence of the researcher in the data collecting process was considered essential (see Appendix 1). With a small sample, it is possible that no rich polar cases would have been included. This would possibly have made the findings less 'rich' in data. Additionally, any research design includes a certain level of arbitrary choice. In large surveys, for example, using a closed questionnaire sent by mail to companies chosen at random, the researcher has to face the choice of *to whom* to send the questionnaires. It is unlikely that any research design for this kind of organizational research would prescribe the choice of the respondents within a company in a random way.

It seems that the researcher was less flexible than he could have been, during the pilot study, in terms of changing the research instrument Had he commenced amending the questionnaire's structure and even challenged the research question itself from the first interview, he could possibly have explored more fully some of the aspects such as the contingent relationships mentioned in the section 'A look forward: some unanswered questions'. It took some time before one of the main advantages of the case studies started to be used, i.e. the possibility of redirecting the research during its course.

Concluding, although the methodology adopted does have problems, it seems that if any other research design had been adopted instead, it would have encountered even more problems, given the objectives established at the outset of this research.

Appendix 1 Methodology issues

Criteria for the choice of research method

The choice of method is particularly important in organizational research. It should ensure that it is possible to address the research problem in a valid way. The method selection should, at the very least, take the following criteria into account: the adequacy for the concepts involved, the adequacy for the objectives of the research, the validity and the reliability.

Adequacy for the concepts involved

The categories, or tentative variables, with which the research is concerned, e.g. environmental uncertainty and manufacturing flexibility, are not concepts which have been sufficiently explored by the literature. For example, the terminology itself is not generally agreed upon (authors are still working on the definition of flexibility); terms such as flexibility and uncertainty are used with several and sometimes different meanings. For that reason, the presence of the researcher during the data collection process is considered essential - therefore limiting the possible size of the sample - to clarify concepts and to ensure that the understanding of the concepts involved is consistent and precise across the subjects.

Another characteristic of the categories involved is the difficulty researchers encounter, when trying to quantify them. For there are no commonly agreed upon objective measures of flexibility or uncertainty in the literature (see chapter 2 for a discussion on these matters). To cope with this difficulty, the researcher should investigate the manager's perception of such variables, rather than the objective measures of the variables. The method should therefore be able to accommodate a perceptual approach.

Adequacy for the objectives of the research

The precise confines of what could be found were not determined at the outset of this research. There was only a literature-based belief that correlations between the perception of uncertainty and the level of variability required and the perception of the need for flexibility of the manufacturing system would be found to be positive. The identification of causal relationships between the categories is central to the present research work. The method chosen should thus also be able to allow for the building of theory regarding these causal relationships.

The main idea is not therefore to verify well established hypotheses but, rather, to build theory. In this sense, a deeper understanding of the functioning of the organization is needed, in order for the researcher to understand the subjects' perceptions and the decision making process regarding the variables involved (Bryman, 1989).

Validity

There are three types of validity:

i) construct validity;

ii) internal validity; and

iii) external validity

Construct validity A method should establish correct operational measures for the concepts being studied to make sure that the information collected actually represents such concepts.

Internal validity It is a concern only for causal and explanatory studies, where an investigator is trying to determine whether event x leads to event y. If the investigator incorrectly concludes that there is a causal relationship between x and y without knowing that some third factor - z - may actually have caused y, the research design has failed to deal with some threat to internal validity (Bryman, 1989).

External validity This deals with the problem of knowing whether the findings of a study can be generalized beyond the immediate case study. This generalization can refer either to the theory involved (analytical generalization) or to the enumeration of the frequencies found (statistical generalization) (Yin, 1988).

Reliability

The objective is to ensure that, if a later investigator follows the same procedures as described by an earlier investigator, the later investigator would arrive at the same findings and conclusions.

Qualitative vs quantitative research: strengths and weaknesses

Quantitative studies tend to give less attention to context than qualitative research. One would not obtain the feel for the organizations under investigation. According to Mintzberg (1979), quantitative studies would not be the most appropriate method to conduct theory building, precisely because 'creative insight seems to require the sense of things - how they feel, smell, seem.'

Quantitative research also tends to deal less well with the processual aspects of the organizational reality. It often entails fairly static analyses in which relationships between variables are explored (Eisenhardt, 1988). In quantitative research, the confines of what can be found are determined at the outset, so that there is rarely an opportunity to change the direction of the research, since the structure largely determines the course of events. An advantage of qualitative research is that it allows for such changes in direction.

The proximity of the qualitative research to organizational phenomena contrasts with the distance between the researcher and the subject that quantitative research normally involves. In field experiments and interviewing, the researcher may have a great deal of contact with the organization, which enables him to develop a fairly strong sense of how it operates. Bryman (1989) points out that qualitative researchers however should also be aware that the proximity, if not well managed, can represent a higher risk of exercising undesired interference with the phenomenon studied.

Critics of the qualitative approach often point to the fact that a qualitative investigator fails to develop a sufficient operational set of measures and that 'subjective' judgments are used to collect data, jeopardizing the construct validity of it. Such a criticism could be overcome, though, by developing multiple sources of evidence to compensate for these deficiencies.

A further criticism of the qualitative approach is the difficulty researchers face when trying to replicate it. The general way of approaching the problem of replication in qualitative research is to make as many steps as operational as possible, and to conduct the research as if someone were always looking over your shoulder (Eisenhardt, 1988). The development of semi-structured protocols is also a tactic to increase the reliability of qualitative research (Yin, 1988). Miles (1979) points out that the quantitative view of reliability is in many respects inapplicable in qualitative data collection: 'Certain cases of reliability must be intentionally violated in order to gain a depth of understanding about the situation i.e. the observer's behaviour must change from subject to subject, unique questions must be asked of different subjects... there is an inherent conflict between validity and reliability - the former is what field work is specially qualified to gain, and increased emphasis on reliability will only undermine that function'.

Macro-approaches to this research work: conclusion

In most of the cases, there is not a free choice of research design[1]. It is in general a matter of appropriateness to the research requirements and conditions. Given the criteria and a brief description of the two distinct approaches for research method - qualitative and quantitative, a matrix criteria vs. alternatives (see Table A1.1) can be established in order to support the method choice:

Table A1.1 Schematic choice of the research method: qualitative vs. quantitative

Criteria	Research main needs	Approach Qualitative	Quantitative
Adequacy for concepts	-presence of researcher in the data collection	usual	unusual
	-small sample size	possible	insufficient
	-variables difficult to quantify	possible	inadequate
	-perceptive measures	possible	difficult
Adequacy for objectives	-confines not pre-defined	possible	impossible
	-causality is central	preferable	possible
	-need to build theory	adequate	inadequate
	-in depth understandin of organizations decision making process	adequate	inadequate
Construct validity		possible	possible
Internal validity		possible	possible
External validity	-generalizable theory	possible	possible
Reliability		possible	possible

The qualitative approach was considered the most appropriate one for this research. In terms of the criteria adequacy for this research (concepts and objectives), in other words, regarding the research requirements, the qualitative approach is clearly superior. With respect to the issues of validity and reliability, one is unable to discriminate between the two approaches, if the research design and data collection process is properly conducted.

The choice of the research design

The principal organizational research designs are, according to Bryman (1989):

i) experimental research;

ii) survey research;

iii) case study/qualitative research; and

iv) active research.

The choice of research design

Experimental research This design was not considered to be suitable for this research on the grounds that it is very difficult to design a representative experiment (or model) which includes variables, such as perceived environmental uncertainty or manufacturing flexibility, in the sense they are considered here. They are very complex and dependent on a number of other complex variables like the subjects' cognitive process; thus any attempts to control them appear to be very difficult.

Survey research More generally, designs based on self-administered questionnaires are not adequate for the present research because:

i) they are methods in which data collection is generally made in the absence of the researcher; and

ii) their main concern is the quantification of variables and enumeration of frequencies.

The presence of the researcher is considered essential in this research, because of the risk of non-homogeneous interpretation of the concepts involved across the subjects. The quantification of the variables and enumeration of frequencies were not considered to be of particular relevance for this present research because the main objective of this research is to build theory rather than to describe the reality in terms of statistical distributions.

Action research It is unsuitable for this research because the researcher was not in the position of suggesting any lines of action to the case-companies.

Case study / qualitative research This seems to be a suitable research design for the present research. For it can, if well administered, provide the adequate level of contact between the researcher and the subjects involved. In addition, it can also provide the appropriate level of detail in the data collection. Case study is also particularly suitable in the case of the lack of previous knowledge about the confines of what could be found in the research and the possible need of redirecting it, should events require this. Moreover, case studies are suitable to allow for the building up of theory (Mintzberg, 1979; Eisenhardt, 1988), which is one of the main aims of the present research.

A summary of the research design alternatives and choice

Table A1.2 summarizes the process of research method selection.

Table A1.2 Summary of the process of research design choice

Research requirements/ characteristics	experimental research	survey research	case study/ qualitative research	action research
-presence of the researcher in data collection	possible	unusual/ difficult	usual	usual
-small sample size	possible	unusual	usual	usual
-variables difficult to quantify	possible	possible	possible	possible
-perceptive measures	possible	possible	possible	possible
-confines not pre-defined	unusual	difficult	adequate	possible
-causality is central	adequate	possible	adequate	possible
-need to build theory; to answer a 'how' question	possible	difficult	adequate	possible
-in depth understanding of decision making process	difficult	difficult	adequate	possible
-non-active role of researcher	possible	possible	possible	impossible
-lack of control over variables	difficult	possible	possible	possible

Overall conclusions on the selection of the research method

From the above analysis, the general approach used in the present research work is predominantly qualitative, and the research design is case studies.

To summarize, what this means is:

i) a number of organizations will be chosen and analyzed in depth. The choice of the organizations will not be made at random. Rather, the criteria for choosing them will be their potential contribution to the theory-building exercise; and

ii) the basic method of data collection will be interviews with a number of decision makers within the organizations in order to identify their perception with regard to a number of aspects related to the research question. A semi-structured questionnaire will be used in the interviews, to be performed by the researcher in person.

See (Corrêa, 1992, chapter 7) for a detailed description of the research design development.

The level of analysis

As Gerwin (1986) points out, a basic aspect in addressing manufacturing flexibility issues is the level of aggregation on which the research is to be based. Gerwin suggests the following classification of levels: the individual machine or manufacturing system; the manufacturing function, such as forming, cutting or assembling; the manufacturing process for a single product or group of related ones; the factory or the company's entire factory system. At each level, says Gerwin, the domain of the concept of flexibility may be different and alternative means of achieving flexibility will therefore be available. Slack (1990a) also addresses the issue of level of analysis. He argues that, from a strategic viewpoint, the most serious oversight in the literature concerns the level of analysis of most treatments of manufacturing flexibility. Slack defines four levels of analysis: the level of the firm, the level of the function (which, not to be confused with Gerwin's definition of function, concerns the manufacturing function as a whole) or total system, the level of the cell or small system and the level of the particular resource.

The underlying assumption of this research is that the primary reason for a company to wish to develop flexibility (or any other manufacturing objective) is to help the organization to compete. In other words, we are particularly interested in the strategic aspect of flexibility. Slack (1990a) points out that system flexibility (which can be understood as a production unit within a plant) would seem to be the most appropriate level of analysis for any examination of strategic flexibility, since it is the system's flexibility (as opposed as the individual resources') which contributes most directly to company's performance.

The level of analysis considered in this research is hence the level of the manufacturing systems, or set of manufacturing resources. This level of analysis does not necessarily encompass the whole factory within companies (which, as in the case of car manufacturers, can sometimes mean huge plants), but it can also apply to relatively independent production units within the plant. Nowadays, with the concept of manufacturing focus being adopted by many companies[2], it does not seem to be appropriate to deal with, or to study, the flexibility of large plants as a whole. Given that frequently, different cells (which may focus on different products or parts) or plants-within-the-plant have different requirements in terms of the performance regarding either flexibility or other competitive criteria.

The important point is that the level of analysis considered here is of relatively autonomous sets of multiple resources (machines, material, people, systems) under common management and not the level of the individual resources or groups of similar resources (such as a lathe or the cutting department in a highly bureaucratic organization).

The choice of the companies

In case studies, cases are not chosen at random. Rather they are selected to fill theoretical categories and polar examples (Eisenhardt, 1988; Pettigrew, 1988; Yin, 1988).

The cases in this research were chosen from companies, in both England and Brazil. The reason for this selection lies in the tentative variables analyzed and also in the possibility of access. The access to English companies was made possible through members of the staff of the Warwick Business School, who had previous contacts with the case-companies. The access to Brazilian companies was possible because of contacts previously established by the author when in Brazil. A split sample was chosen for the following reason: the industrial environment in Brazil is notoriously more uncertain than the industrial environment in England. Following Pettigrew's (1988) advice, it was decided that it would make 'pragmatic sense' to choose such an extreme situation which would allow the analysis of a very uncertain environment. However, because the Brazilian industry has been protected from foreign competition for a long time it is not as developed as the English industry, in terms of product proliferation. Consequently, English companies were thought to be more apt at providing good data for valuable analyses in terms of variability of outputs. Thus, with companies from both countries in the sample, both variables - uncertainty and variability - could be analyzed based on polar cases.

The Brazil/UK factor

The non-uniformity of the sample, in terms of the countries the companies are located in, was not considered a methodological problem for two reasons.

Firstly, the sample is not intended to be representative of a specific population. From the outset of the research work, no statistical generalization was intended. [3] Secondly, from an operations viewpoint, the problems which a company belonging to the automotive industry face are of a similar nature, be it located in Brazil or in the UK. For the *hard* part of the processes are similar, e.g. the machines or the assembly operations, although the uncertainty regarding them is probably different. In terms of the 'soft' part of the process, the organization, systems, and so on, the case-companies in both countries are still similar; of the two Brazilian companies in the sample, one is part of a large multinational group with headquarters in Europe and the other, because it is highly export-orientated, having to meet European and American standards rather than simply Brazilian ones, also follows European and American models of production organization and management. An alternative approach would have been to keep the whole sample either totally Brazilian or English, but, in doing so, the richness of the extreme cases would be lost.

The case studies were done based on semi-structured interviews with a number of managers within the organizations.

The number of people interviewed varied from company to company, depending on the specific organizational structure, on their availability and on their willingness to cooperate.

The research instrument

The case studies in the present research work were based on semi-structured interviews. The interviews were conducted personally, on site, by the researcher for the reasons discussed previously. An instrument (or protocol) was developed to be used in the interviews for the following reasons.

Firstly, it operates as an *aide-memoire*, ensuring that the same aspects are covered with all the subjects, thus increasing the reliability of the research.

Secondly,it helps focus the process of data collection on the relevant issues.

Thirdly, it makes it easier to register the information collected, given that some of the subjects might have preferred not to have the interviews tape-recorded.

The design of the first version of the protocol

When first designing the protocol, one of the main concerns was exactly what questions or topics to include. This was because the subjects should not be overly influenced, although in any classification structure or question sequence there is always some sort of presupposed viewpoint[4]. On the one hand it was clear that the subjects could not be approached without any instrument or with very open questions, since the amount of time to spend with them was limited and consequently the need to focus on determined issues was very important. On the other hand the subjects should be given the opportunity to go beyond the strict confines of closed questions. It was intended to identify their views on the issues, as opposed to having them simply confirming or denying the researcher's a priori views.

Initially, it was decided to have the protocol divided into sections, each contemplating a different tentative variable: environmental uncertainty, variability of outputs, flexibility of the resources and flexibility of the manufacturing system. It was decided to address flexibility through both levels - that of the individual resources and that of the system of resources. The reason for this decision if that, on the one hand the level of analysis chosen is that of the system of multiple resources and we expected many of the uncertainties (such as uncertainties related to the demand) to be related to the system as a whole, thus possibly calling for actions at the system's level. On the other hand as Slack (1987) suggested, based on his empirical findings, managers feel more comfortable when talking about the flexibility of the individual resources. In light of this, it was considered that approaching the issue of the flexibility of the resources (as well as the flexibility of the system) would increase the construct validity of the research. At the same time, we expected that some of the uncertainties could be related to specific resources (such as machine breakdowns), possibly calling for actions at the individual resource level, although affecting (and therefore being of interest for the analysis of) the system as a whole.

The main concern of this field work was to attain an understanding of the managers' perceptions with regard to the relationships between the three tentative variables - environmental uncertainty, variability of outputs and manufacturing flexibility. The intention was to analyze the relationships from the managers' perspective. With this in mind, it was thought that, if the researcher adopted the approach of asking the managers to confirm or refute the a priori relationships found in the literature, the possibility would arise of inducing the managers to accept the a priori ideas. This would increase the risk of jeopardizing the internal reliability of the research. For this reason, it was decided to treat each different section as a self-contained part. In other words during the interviews we would talk with the managers about each of the tentative variables involved in the research, namely uncertainty, variability of outputs and manufacturing flexibility, separately. For each of them, we would try to identify the manager's perceptions about a number of different aspects or *factors* taken from the literature, from the researcher's past experience, and from logical thinking. For each of the aspects covered, we would ask about the level of the variable (such as the level of perceived uncertainty regarding the aspect 'machine breakdowns', for instance); yet, we would also ask how important the managers regard the aspects to be to their operation.[5] To allow the managers to scale their answers, and to facilitate comparisons, 5-point Likert scales[6] were used whenever the managers were asked to state their perception.

The 'uncertainty' part of the protocol Factors which are possible sources of uncertainties were listed and subjects were asked to comment on and grade their perceived levels of uncertainty regarding each aspect, using a 5-point Likert scale varying from not predictable to completely predictable. They were also asked to comment on and grade the aspects' importance with regard to the operation's functioning, using a 5-point scale varying from 'not important' (point 1) to 'important' (point 5) .[7]

The 'variability' part of the protocol The variability part of the questionnaire was designed to be a more objective one. Questions were asked about the families of products and their components (how different the products within a family are), the contribution of each family to the sales turnover, the number of different products within each family and the importance (from 'very important' to 'irrelevant') of having that variety for the company's competitiveness. The difference between products within a family was assessed using the following 5-point scale: 1. single product; 2. minor differences between products (colours, accessories); 3. fairly different products made-to-stock; 4. assembly to order (but make to stock) according to customer's specifications; and 5. completely different products made to order according to customer's specifications.

The (first) 'flexibility' part of the protocol (flexibility of the manufacturing resources) Aspects regarding the individual resources drawn from the literature which were considered relevant to the system's flexibility were listed, and the subjects were then asked to comment on and assess:

i) the performance of the company with respect to each of the aspects (the 5-point scale varying from 'very good' to 'very bad'); and
ii) the importance the subjects attributed to these aspects for the organization's competitiveness (on a scaling from 'very important' to 'irrelevant').

The (second) 'flexibility' part of the protocol (flexibility of the manufacturing system) The assessment of the perception of the managers regarding the flexibility of the manufacturing system as a whole was made by using Slack's (1989) model. Slack's classification was selected, since his was considered the most consistent of all the classifications found in the literature, given the level of analysis we intended to develop in this research. The other classifications available either mixed different levels of analysis (e.g. Browne et al., 1983) or approached flexibility at a different level than the one adopted here.

The subjects were asked to comment on and assess the system's performance, compared to that of their main competitors. For this, they were given Slack's (1989) 4 types - product, mix, volume and delivery - and his two dimensions - range and response - of the manufacturing system flexibility (scaling varying from 'much better' to 'much worse' than the competition). They were also asked to assess the importance they assigned to each of the factors in terms of the competitiveness of the organization (the scale varying from 'very important' to 'irrelevant').

Perception of performance vs. perception of importance

The main reason for including the assessment of the perceived performance of the aforementioned aspects in this research was to give the subjects the chance to clarify the concepts involved, by commenting on them and going into the subject in more detail. Based on the comments received, the researcher was also given the chance to ensure that the subjects did not interpret the terms in a way other than the intended meaning[8]. The information about the managers' perception of performance, which was collected, was not however analyzed further, since what mattered most for the present research was their perception of the *importance* of the factors, i.e. what the managers regarded as being the *best way of doing things* rather than the way things are currently being done. For the way things are actually being done can be affected by several circumstantial problems, such as cash-flow restrictions. We were more interested in the longer term view of the subjects, which was seen to be more appropriate for the theory building exercise. Another reason for not considering the perceived performance was that frequently the subjects did not have enough information on what the performance of some aspects actually was like. This lack of information was due, mainly, to the level of decision and span of control of the managers interviewed. (In a number of cases the plant managers did not have information about the performance of the competitors, for instance.) In these circumstances, what the research aims at is to identify the manager's preferred solutions rather than the actual performance of the organizations, regarding the variables analyzed. This can be schematically seen in Figure A1.1.

The ordering of the sections

It was a deliberate choice to design the questionnaire with the sections in the following order: 1. uncertainty; 2. variability; 3. and 4., flexibility.The reason for this was that we did not want to talk about possible *solutions* (the flexibility aspects), before talking about the *problems* (the uncertainty and variability aspects). In structuring the questionnaire in this manner, we avoided inducing the subjects to feel compelled to point to the solutions suggested by the literature.

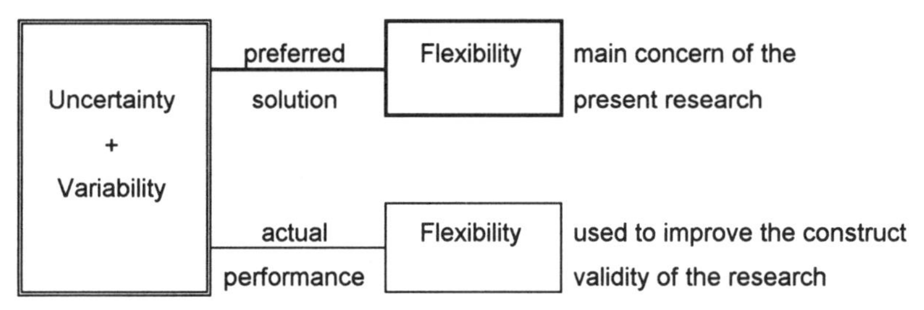

Figure A1.1 Perception of importance vs. perception of performance

The refinement of the research instrument: the pilot study

A pilot study was conducted with four companies, one in England and three in Brazil, the aim being to refine the research instrument. One manager was interviewed in the English company and three managers in each of the Brazilian companies, all of them with a span of attributions which was consistent with the level of analysis of this research. The four companies of the pilot study are suppliers of metal engineering parts for the automotive industry and leaders in their markets. They are briefly described below:

Company X/UK An English company, part of a large corporation, which manufactures mechanical parts for the automotive industry and also parts for the process industry machinery. One person was interviewed - the production director.

Company X/Brazil: The Brazilian branch of Company X/UK. Company X/Brazil has a narrower line of products than the one manufactured by the British branch and aims basically at the automotive industry. Three people were interviewed - the managing director, the industrial director and the production manager.

Company Y: A manufacturer of high quality parts for the automotive and aero space industry. Company Y has the largest milling shop in Brazil. Three people were interviewed - the industrial director, the managing director (who was a former industrial director) and the process manager.

Company Z: The largest Brazilian manufacturer of off-road military heavy and light vehicles, a major exporter of weaponry and off road military vehicles. Three people were interviewed in the pilot study - the managing director, the quality manager and the materials supply director.

The instrument was then modified substantially from the first version, based on the process and on the outcomes of the pilot study. The main problems encountered with the first version were:

Internal validity

After interviewing the first three companies in the pilot sample, it became clear that it would be impossible to establish causal relationships between the variables from the data alone, if the managers continued to be asked only about their perception of level of performance and importance of the three variables - uncertainty, variability and flexibility, in isolation. The fact that a manager, for instance, considers the uncertainty factor UF-X to be highly uncertain and highly important and at the same time considers the resource flexibility factor RFF-X as highly important does not necessarily imply that he considers RFF-X important for coping with UF-X. Both the uncertainty factor and the flexibility factor could actually be 'numerically' correlated, without any causal relationship at all. Probably a much larger sample would be necessary for the researcher to be able to conclude anything about causality with a reasonable level of confidence, using such a procedure. In view of this problem, it was decided to change some aspects of the data collecting process to be able to establish causality, and therefore increase the internal validity of the instrument:

Uncertainty part Instead of simply enquiring about the levels of uncertainty (or predictability - see chapter 3) of the factors and their importance, it was decided, first, to continue asking about the levels of uncertainty of each factor. Second, instead of just asking for levels of importance, it was decided to ask the managers to rank the uncertainty factors which they regarded as representing the higher levels of risk for their organization's competitiveness.[9] The managers were also permitted to mention as many uncertainty factors as they wanted. Once the factors were ranked by the managers, they would then be asked what was, according to their perception, the best way to cope with each of them. These could be ways which were actually being used, if the managers were satisfied with them, or ways which they considered should be used by their organizations to cope with the risky uncertainty factors. It was expected that answers would relate to flexibility factors, but, by having the uncertainty part of the protocol first, it was made sure that the subjects were not induced to answer the factors listed in the flexibility list, given that they had yet to go through that section.

Variability part It was decided to include a question at the end of the variability section: the managers were asked by the researcher what they considered would be the best way for their organization to cope with the variability of outputs which they have to deal with. Although answers connected to flexibility-related factors were expected, once again it was made sure that the subjects had not been induced to answer according to the listed flexibility factors.

Flexibility part The resource flexibility part was amended in a similar manner to the uncertainty part. Instead of simply asking the subjects to assess the levels of performance and importance of each of the factors, it was decided to, firstly, continue asking them to go through the whole list of factors, commenting on and also ranking their performance (compared to the company's needs). Thus it was ensured that they continued to have the opportunity to go into the subject in greater detail and also the researcher was given the opportunity to make sure that the subjects' understanding of the concepts was appropriate. Secondly, instead of asking the subjects to grade the performance of all the aspects, it was decided to ask them to rank the listed factors (although they were permitted to mention others), according to the ones they considered as being critical success factors[10] for the organizations' competitiveness. It was up to them to decide the number of factors they mentioned in the ranking. After ranking the factors, they would be asked to comment on why they considered the ranked factors to be important. At this stage, we expected to find answers relating to either uncertainty factors or variability factors.

The way of getting the questions to the respondents

It was noticed that, if the questionnaires were sent to the managers in advance this could induce the managers to answer according to the a priori classification, since they were given the opportunity to go through the list of solutions (or the flexibility factors) before answering what is the best way of dealing with the problems. This then might cause them to answer according to the list and not according to their own views. To avoid the possibility of a bias of this nature arising, it was decided not to send them the protocol in advance. The protocol would be presented to the interviewees at the beginning of the interviews.

The incompleteness of the lists

It was noticed that our list of factors for each of the variables could be restrictive. Thus, for example, at the end of the sections on uncertainty and flexibility of resources, there came a section where the subjects were asked whether there were other factors which they considered important, but which had not been included in the list. With the inclusion of this section, it would be possible to take into account from then on (in the current and in the following interviews) any other relevant aspects not mentioned before. As a result of this action, by the time the pilot study was finished, the protocol was found to be substantially different from its initial format. This inclusion, moreover, proved to be an instrument for the fine tuning of the protocol, which had little alterations made to it during the actual process of data collecting. For although, only minor changes were made to the list of questions during the process of data collection itself, substantial changes in the data collection methods are also legitimate when one conducts case studies for theory building. According to Eisenhardt (1988):

> Indeed, a key feature of theory building, case research is the freedom to make adjustments during the data collection process. They can be changes to data collection instruments, such as the addition of questions to an interview protocol or questions to a questionnaire.

The inconsistency of the questions about 'importance'

It became apparent that inconsistent and possibly dubious terms were being used, when asking the managers about their perception of the 'importance' of the factors. The most appropriate way to obtain a greater degree of consistency was to link the question of importance to the question of competitiveness of the organization. So, instead of talking about the 'importance' of the uncertainty factors, we talked about the 'risk for the organization's competitiveness'. Instead of talking about the 'importance' of the flexibility related resource factors, we talked about the 'critical success factors to the organization's competitiveness'. The system flexibility section remained as in its initial format, since the question asked to the managers in that section already mentioned the importance of the aspects for the organization's competitiveness.

The inclusion of other aspects of variability

Instead of asking only questions regarding the variety of the product line, we included some questions regarding the variation in quantity of the outputs, such as percent variation of overall volume, so that the managers did not feel constrained to talk simply about the variety of products.

Who and how many people to talk to

The managers interviewed were chosen at the discretion of the researcher, among a list of possible interviewees, drawn from the initial contact with each company. In general, this first contact was the contact in which the researcher negotiated and gained access to the company. The decision among the alternative managers was made mainly on the basis of who were the people considered to make decisions concerning a set of multiple manufacturing resources and also considered to have relevant experience and information about the company to be able to give the researcher a richer picture in the interviews. Experienced production managers, process managers, quality/productivity managers, semi-autonomous manufacturing cell leaders are some examples of subjects considered appropriate for the purposes of this research. In contrast, dedicated machine operators, cost accountants, design engineers, traditional marketing managers, finance managers and also some CEO's are examples of subjects which were considered inappropriate and therefore not selected to take part in the interviewing process. The number of people interviewed varied from company to company depending on the specific organizational structure and on their availability and willingness to cooperate.

The treatment of the data

According to Eisenhardt (1988), analyzing data is the heart of building theory from case studies. It is both the most difficult and the least codified part of the process. Eisenhardt suggests the following key steps for case analysis:

Within case analysis Generally, it involves detailed case study write-ups for each site. The need for within case analysis is driven by one of the realities of case study research: a staggering volume of data and therefore the ever present danger of 'death by data asphyxiation' (Pettigrew, 1988). Within case analysis typically involves detailed case study write-ups for each site (Eisenhardt, 1988) and helps the researcher to start the process of progressively making sense out of the large amount of collected data.

Cross case search for patterns The idea is to force the investigator to go beyond initial impressions. The danger here is that investigators reach premature and even false conclusions. One tactic is to select categories or dimensions and look for similarities coupled with inter-group differences.

Shaping hypothesis The central idea is that from within case analyses, plus various cross site analyses and overall impressions, tentative themes, concepts and possibly even relationships between variables begin to emerge.

Following the suggestions of the literature, efforts were made to ensure that the treatment of the data from the cases was as systematic as possible. The first step was to listen carefully to the tapes and transcribe literally every relevant bit of the interviews. The next step was then to highlight successively the pieces which represented the relevant relationships for the research, sifting information progressively by making successive passes, until the relationships between key words that were representative of the factors were identified[11]. Elaborate charts were then constructed which summarized the results and consolidated the interviews from each company. These charts would bring together a summary of the relevant information and the 'addresses' of the detailed information, making references to the transcriptions, to the notes taken during the interviews and to the tapes. In using such a procedure, one could always clarify any doubts about the summary results by tracking them back to the primary data. In a number of situations, during the process of treatment of the data, some interviewees had to be contacted again, in order to supplement missing information.

The within case analysis

From this stage, detailed within case analyses were completed and the results written up. The next step was to conduct the cross case analyses, searching for patterns, relationships, similarities and differences between cases. All the case write-ups are structured in a similar way. The following sections can generally be found in each case write-up, because either they refer to relevant general information on the case companies or they are directly related to the research question and propositions:

Organizational issues This section contains a brief description of some relevant organizational aspects of the companies. The objective is to place the case in context.

The interviews The managers interviewed are mentioned, together with a brief description of each of their responsibilities and activities.

Line of products: variety, variation and innovation Aspects regarding the variability of outputs of the company are discussed.

Manufacturing flexibility task and performance Aspects regarding the respondents' perceptions of the manufacturing system's flexibility are discussed.

Uncertainties involved The respondents' perceptions of the uncertainties (associated with predictability) involved in their operations are discussed.

Coping with change, uncertainties and variability According to the managers' perceptions, the relationships between aspects of the three main research variables - uncertainty, variability and flexibility - are discussed.

The relationship between the flexibility-related order winning criteria and the critical success factors Aspects related to the relationship between the system's flexibility factors and the resource flexibility factors are discussed.

Conclusions of the within case study Conclusions of the within case studies are drawn.

The cross case analysis

The next step was to conduct the cross-case analysis, trying to search for patterns. An analysis regarding the similarities and differences was conducted, conclusions were drawn and comparisons with the research propositions were made. The result of the data treatment process can be found, summarized, in Appendix 2.

Based on the results of the case study, an attempt to build theory was then made, the results of which are presented in chapter 4.

Brief summary of the method used in the research

i) Qualitative research as the general approach.
ii) Case study as the research design.
iii) Semi-structured interviews as the basic data collecting method.
iv) Emphasis on the perceptions of the decision-makers with regard to the research question and propositions.
v) Production units (factories or semi-autonomous parts of factories) as the level of analysis.
vi) Four organizations, two in England and two in Brazil, belonging to the automotive industry, manufacturers of metal engineering products, as the case-companies, besides the four companies of the pilot study.

Notes

1 According to Morgan and Smircich (1980), the appropriateness of qualitative research - like that of quantitative research - is contingent on the nature of the phenomena to be studied.

2 Semi-autonomous production units within plants are frequent nowadays, with the companies adopting the focused manufacturing and 'plant-within-a-plant' approaches (see chapter 1 for a discussion on the issue).

3 Case studies rely on analytical generalization rather than statistical generalization as is the case with survey research (Yin, 1988).

4 'What is more, observation as such cannot be prior to theory as such, since some theory is presupposed by any observation... Twenty-five years ago I tried to bring home the same point to a group of physics students in Vienna by beginning a lecture with the following instructions: "Take a pencil and paper; carefully observe, and write down what you have observed!" They asked,. of course, what I wanted them to observe. Clearly the instruction: "Observe!" is absurd... Observation is always selective. It needs a chosen object, a definite task, an interest, a point of view, a problem. And its description pressuposes a descriptive language, with property words; it presupposes similarity and classification, which in turn presupposes interests, points of view and problems.' (Popper, 1972, refered to in Magee, 1990).

5 We expected initially that by simply obtaining their perceptions of the importance of the several aspects of each variable for their operation, we would be able to draw correlations between the variables. During the running of the pilot study, this eventually proved to be a mistaken assumption.

6 Likert scales are now widely used in assessing people's perceptions. They were first used by R. Likert (Likert, 1967), in order to study the informal structures within organizations. Likert scales are continuums representing certain aspects of the organization, upon which responses can be made. The continuums are divided into intervals. In completing the questionnaire, or answering questions in an interview, an individual is asked to place a mark on the continuum at the point which best describes his/her perception about the particular aspect of the organization under investigation.

7 Eventually we changed the way we treated the 'importance' aspect and also the way we defined it.

8 This was done to ensure the construct validity of the instrument.

9 Ranking the factors instead of assessing the importance of all of them also substantially reduced the time of the interviews. This was an additional advantage, since the interviews were found to be too long in the first interviews.

10 Critical success factors were defined for the managers as the factors which they considered the most important for the competitive success of their companies.
11 This is what Miles (1979) calls 'formulating classes of phenomena', which is essentially a categorizing process, subsuming observations under 'progressively more abstract concepts'.

Appendix 2 Case studies

Introduction

As described in Appendix 1, the research method used in this work is case studies. The research problem is defined as 'the relationship between manufacturing flexibility, variability of outputs and environmental uncertainty'. Some information about the protocol used in the interviews with the managers as well as the method used to collect and analyze the data presented in this chapter are also briefly described in Appendix 1.

The case-companies

Four plants were analyzed two being in Brazil and two in the United Kingdom. All of them are manufacturers of engineering products and all of them belong to the automotive industry. One of them is a vehicle manufacturer and three of them are vehicle sub-assembly manufacturers, one manufacturing carburettors, the other, engines and the last one, shock absorbers.

The within-case analyses

Case A - the English engine plant (Company A)

Company A is an automobile manufacturer located in the Midlands, England manufacturing to stock and assembling to order. This case relates to the engines manufacturing plant within Company A. Some figures about Company A are presented below.

Number of Employees	
Direct	610
Indirect	278
Number of products/week	
4 cylinder - 32 derivatives	940
V-8 - 46 derivatives	440
approx % of components made in (in number)	40%
approx % of components bought in (in number)	60%
approx % of components made (in value)	60%
approx % of components bought in (in value)	40%

Organizational issues Company A's engine manufacturing plant is organized in manufacturing cells by manufacturing task. There are eight main cells each of them with one manager, one facilitator, one planner, between one and four conformance engineers and between 12 and 120 direct workers. The facilities layout of the cells does not always follow the usual U-shape or circular shape suggested by many authors in the literature. In some cases, the term 'cell' is broadly used to define a sector of the plant under the same manager. Although focused on a defined manufacturing task and in general, organized on a product rather than on a process basis, the cells are not always able to manufacture a complete part or component; yet this was mentioned as a goal by the managers. The cells' management has considerable autonomy in deciding on scheduling and dispatching, employment, training and, to a more limited extent, on investment budgets. There are statistical process control procedures implemented and the workers are responsible for the process quality. Maintenance is still performed by a separate team, although it is intended to be delegated to the operators and cell managers in the future. This arrangement also applies to the setting up of the machines.

The formal manufacturing planning and control system is basically MRP[1] for the planning of materials and master scheduling. The scheduling within each cell however is done on a people-based informal system by the cell manager and staff.

The engine plant works under a production director with a staff of one conformance manager and his team, who are responsible for manufacturing engineering, methods and quality control (precision components, final product tests).

The general approach regarding industrial relations has recently changed towards more stable relations. In the words of the production director:

> [regarding the variation of overall demand volume] ...the old choice is to drop the line rate and shed the labour. We now say no. We have a lot of

training, a lot of money invested on them, let's keep them. So we keep the rates and on Friday morning we deploy everybody, from the four corners of the Company; training, quality action teams, and we are actually making people better. And we keep them. This is a major change in views. Now we say: 'We may not need to produce vehicles now, but we still need you. We need your brains'

The payment system for direct workers is based on four grades, according to the breadth of skills of the worker.

The interviews Six people were interviewed in Company A:

The materials manager, referred to in this case as 'Manager 1', responsible for the material flow management: receipt, handling, storing and moving.

The conformance manager, referred to as 'Manager 2', responsible for finished product quality control, testing and industrial engineering

The CNC cell manager, referred to as 'Manager 3', responsible for the cells which manufacture a number of different parts made of aluminium or steel, using CNC machines

The assembly lines manager, referred to as 'Manager 4', responsible for the whole assembly operation of engines

The transfer line manager, referred to as 'Manager 5', responsible for the milling and finishing of crank shafts and engine blocks, performed in two transfer lines

The production director, referred to as 'Manager 6', responsible for the whole production of power train units in Company A

Line of products: variety, variation and innovation Because of the cellular manufacture in Company A's plant, for some analyses, it is more appropriate to consider product variability issues in terms of the particular work units or cells since they vary considerably according to the manufacturing task of each cell. The crank shaft and blocks cell, for instance, has a low variety of products (only four different engine blocks and four different crank shafts) because the parts it manufactures are standardized building blocks and therefore common parts for a number of derivatives. On the other hand the assembly line cell (which actually comprises two assembly lines - one, a simple conventional straight track which assembles for cylinder engines, and another, a serpentine-type line based on AGV's, or automatically guided vehicles, has a much greater variety with 78 different derivatives. The third cell studied, focused on milling steel and aluminium parts using CNC and general purpose machines, has an intermediate level of variety with approximately 65 products divided into two cells - one for pulleys and fly wheels which is not greatly different from each other and which includes a robot to feed two CNC machines and another one for fairly different engine parts made to stock.

The variation in overall volume for the engine plant can be approximately 20 per cent from month to month. The variation in the mix of products demanded can be very high. In each week 45 out of the 78 derivatives are produced, in average. Moreover, every month at least 80 per cent (or approximately 62) of the product range is produced, the remaining 20 per cent being service parts for replacement of phased out vehicles or special low volume orders.

The introduction of new products or engineering changes is done on a batch basis, quarterly (or every 13 weeks). In general, around 15 changes are performed in each batch. Five are generally substantial, with changes in the process, and ten are minor changes in the assembly line. A totally new product or a completely new derivative appears every year or every other year.

The process of launching a completely new product is briefly described by the production director as below. The description is based on a real engine development which was part of a new vehicle which was completely designed, developed and launched in 3 years:

> If we are introducing a brand new engine, basically, first of all it is defined in terms of in which vehicle it is going. The big issue nowadays about engines is emissions controls. We are working on a lot of things like that, to make an engine which is cleaner. You measure in that, probably, a period of two years from the initial concept to when we actually start to make bits and prototypes. The director of product and process will have people there. (Manager 6)

Recent changes in organizational structure resulted in process and product development being combined under the same director, in order to improve integration and ensure simultaneous engineering.

> For instance, now I'm sitting alongside this same guy, we are working on the same parts, he is developing our engines for 1994, four years ahead, and the beauty is now we can be talking about something, because I'm the manufacturing man, I can say: hang on it, don't do that because if you do that, we'll have a problem. We try to make sure that they design for manufacturing. So that concept starts, the period of time for the 'X' engine (for the new vehicle 'D') was about three years from when we thought we want a new vehicle, what are we gonna do to when we started producing parts for prototype units.
>
> We actually as a company form teams of people, from all over the place, then we put them in this team, they work on the project, when it's ready to launch they go back to their original jobs. When another new one comes on we form another team, usually with different people. So, we formed the

vehicle 'D' team and it was right across the broad span from finance people, designers, product engineers, stress analysis people, technicians, manufacturing people, 12 people (the hard core) managed by the project director for vehicle 'D'. Some 18 months before the final prototypes were decided on, manufacturing actually started building it, then, just then, right from me here there is a secure area, a prototype area. Then we, as manufacturing people, actually helped build the prototypes.

That is a very powerful thing because what we do there is, every six months, we get 2 or 3 good guys off the tracks, and put them in this area and then they work with the project, and then, they fit the new parts and things like that. We then rotate people so they are involved in the new project early, to utilize their local knowledge to modify the design and make it easier to build. We are working now on our engines for 1993 and 1994. Right now it is clear what we are going to do. What we do now, this team of people, remember there is a hard core of the team, and we put more people in the team depending on the stage of the project. Very flexible thing, with people moving to and from the team to their normal jobs. At a certain stage, probably 18 month to 2 years before prototypes, very early, suppliers are involved. What we are moving towards now is we are getting similar to what the Japanese lean production companies are doing. We say: that is the specification of the component we need, can you go and design it for us. And the company then owns the product. We want them to design it for us and work with us. So, you are the specialist, you design it for us. Company A traditionally designed and specified probably 90 per cent of the bits and pieces and my view is that you've got to rely on the experts.

A good relationship is necessary between the company and its suppliers. This is a very long process, to develop this close relationship. This requires a fundamental change in the relationship with suppliers. You don't do that overnight. It takes many years. The first step is to reduce the number of suppliers. (we reduced the number of suppliers from 1200 to about 700) In many important areas we are single source supplied. You are a supplier, instead of a 2 year contract now you have a 5 year contract, and you get it right.' (Manager 6, 1990)

A number of important points can be underlined by the brief description of the process of launching a new product in Company A:

i) integration between process and product development and production via organizational links (such as product and process development functions under one director to ensure simultaneous engineering) or via effective inter-function communication (such as the product/process director and the production director communicating to ensure manufacturability of the products);

ii) multi functional team approach;

iii) early involvement of direct workers in the design and prototyping phases;

iv) early involvement of suppliers and delegation to the experts of the task of designing and developing the parts; and

v) reduction of the number of suppliers and tendency to establish long term contract

Manufacturing flexibility task and performance Although belonging to quite different manufacturing cells, the managers' answers are clear and consistent, concerning the flexibility-related task of the engine plant. All the respondents placed mix flexibility as their first priority. All but one of them specified mix response and one specified mix range as the particular mix flexibility dimension the engine plant should focus on. The next most mentioned flexibility type was volume flexibility. Table A.1. shows the distribution of answers. The figures under the Manager's columns represent the priority each of them gave to the individual flexibility types in terms of order-winning criteria.

'4' means that the manager gave the specific flexibility type first priority, '3' second priority, '2' third and '1' fourth priority or lower. The column 'Total' sums up the total number of managers who mentioned each of the flexibility types. The column 'Total weighed' shows a weighed total considering different weights for different priorities. 4 'points' are associated with first priority, 3 with second priority, 2 with third priority and 1 with fourth priority. So, for instance, product response mentioned by Managers 4 and 5. The column 'Total' therefore shows 2. The column 'Total weighed' shows 2 (3rd priority according to Manager 4) plus 3 (2nd priority for Manager 5) equals 5.

Table A.1 Priority given by managers of Company A regarding types and dimensions of manufacturing flexibility

	Mgr 1	Mgr 2	Mgr 3	Mgr 4	Mgr 5	Mgr 6	Total	Total weighed
Product range				1			1	1
Product response				2	3		2	5
Mix range		2				4	2	6
Mix response		4	4	4	4	2	5	18
Volume range			3	3		1	3	6
Volume response					2	3	2	5
Delivery range							0	0
Delivery response		3	2				2	5

Uncertainties involved The uncertainties mentioned by the managers as the ones which represent the highest potential risk to Company A's competitiveness show a distinct pattern. All the managers, for instance, placed materials and parts supply uncertainty as one of their two main concerns. Three of the managers placed demand product mix uncertainty among their two main concerns, two managers placed labour behaviour - absenteeism and continuity - among their two main concerns. Other perceived uncertainties in Company A can be found in Table A.2.

Table A.2 Importance given by managers of Company A regarding sources of uncertainty as potentially jeopardizing for competitiveness

	Mgr 1	Mgr 2	Mgr 3	Mgr 4	Mgr 5	Mgr 6	Total	Total weighed
parts supply	4	4	3	4	3	4	6	22
labour behaviour	3	1	2	3			4	9
machine breakdown	1	2		2			3	6
systems accuracy		1					1	1
product introductio	1		1	1	1		4	4
mix changes	2	3	4	1	4		5	14
volume changes	1		1		1		3	3
general uncertaint			1	1	1	1	4	4

The figures under the columns 'Mgr' represent their perception on the ranking of risky factors to the organization competitiveness. 4 means first ranked, 3 second, 2 third and 1 forth or lower. The managers were free to choose as many factors as they wished to mention. The column 'Total' represents the total number of managers who mentioned the factor. The column 'Total weighed' sums up the number of times each manager mentioned each factor but considers different weights for different priorities. 1st ranked is associated with 4 points, 2nd ranked with 3, 3rd with 2 and 4th and lower with 1.

Coping with change, uncertainties and variabilities When asked how they coped with the different types of uncertainty and variability, the managers showed, in general, both different approaches and different levels of understanding of the variables involved with manufacturing flexibility. The manager responsible for the CNC machines cell not surprisingly showed a higher level of concern about the issue than the manager of the transfer line cell. Some of the managers were able to classify types of uncertainty and discriminate different types of action which would be necessary to cope with them. Others, on the other hand felt more comfortable talking about 'general uncertainty of the process', suggesting, accordingly, general aspects of flexibility to cope with it.

It is important to mention a certain hierarchy in the general approach adopted by a number of managers in terms of the ways they find the most appropriate for dealing with uncertainties. They generally seem to prefer attempting to reduce the level of uncertainty they suffer rather than to deal with its effects. This can be noticed in a number of situations such as the following:

Manager 6, referring to machine downtime, prefers to control the occurrence of machine breakdowns by means of developing a preventive maintenance system than to have to deal with it by having excess inventory or capacity:

> ... at the moment, we deal with it by carrying high inventory. In certain areas, we keep excess capacity. But the way we intend to cope with it in the future is total preventive maintenance' (Manager 6)

Later, talking about the frequent and unexpected changes in schedule caused by problems in the paint shop, which is physically remote from the engine shop, Manager 6 prefers to control the change by reducing its uncertainty via coordination between his shop and the origin of the change itself:

> ... at the moment we deal with it through the ability of people to react quickly and reschedule. What we are now moving into is when a body is launched, when it is determined, when that happens, if you can have that information at that stage, probably one or two days in front of my process, we know what vehicle is gonna get out of that line. We actually bought some computer equipment which talk to my own equipment directly: when

> a vehicle is launched, they can tell me exactly what that unit is requiring one or two days in front. (Manager 6)

Manager 6 now talks about keeping up with all the technological change which has happened in recent years. He describes a situation where Company A decided to reduce the amount of technological change they would have to deal with by subcontracting a contractor to do it for them:

> Company A traditionally designed and specified probably 90% of all bits and pieces... What we are moving towards now is we are getting similar to what the Japanese lean production companies are doing. We say that is the specification of the components we need, can you go and design it for us. They are the specialists and can keep up with the changes in the engine technology... (Manager 6)

Manager 2, talking about manufacturing flexibility in general:

> Because of the way the site runs, without having a huge amount of resources or a huge amount of inventory, you have to have a certain amount of flexibility. We got to the point that we don't like it, but we are good at it. It would be nice to have it done in a more controlled fashion. (Manager 2)

It seems therefore that Manager 2 also prefers to control and reduce the reasons to be flexible. Nevertheless, he does not always see very clearly the relationship between reducing the change and having flexibility, as can be noticed by his words:

> Being flexible eliminates the pain of not having the system in place. If you had the perfect information, it would eliminate all the need for flexibility. (Manager 2)

This does not appear to be totally valid. If a plant decides to have a very variable output and if, at the same time, it has scarce resources, despite such a demand being perfectly known, the firm will need to have some sort of manufacturing system flexibility developed in order to cope with it.

The words of Manager 6 seem to agree with the preference for the uncertainty reduction, when commenting on the uncertainties they face with the supply of material and parts:

> ...if it is a short term problem, we can change our schedule to accommodate it... We work more in terms of reducing uncertainty, but the inevitable happens and we have to cope. (Manager 6)

and also when talking about standardization, which is a way to reduce the overall number of changeovers:

> ... standardization of parts is important because it reduces the need to be flexible. (Manager 6)

Thus, managers in Company A generally see flexibility as a way to deal with change when its causes can not be eliminated.

The most frequently mentioned relationships between types of uncertainty, variability and ways to deal with them are shown below, together with some representative quotations regarding them. The first column represents the number of managers who mentioned the relationship. The second and third columns represent the relationship itself, the second being the source of the uncertainty and the third, the capabilities or ways managers see as worth developing to be able to cope with the uncertainty shown in the second column.

Number of managers who mentioned the relationship	Uncertainty relating to...	⇒	Best way to cope with it is by developing...
4	parts and materials supply		rescheduling capability

> ...we tend to resequence to work with what we've got or more and more we tend to if possible, if the component is the sort of component which stands alone, or in other words if we can fit it afterwards, we build without that component. If we can resequence, fine. If we can build
>
> without it, fine. If the effect of that is so severe that we can't do that then we stop the line. (Manager 4)

4	labour behaviour (absent, continuity)		multi skills of labour, transferability

... if there is uncertainty with labour, the labour force has to be flexible enough to be able to move around and they have the flexibility of skills to move around within certain limits, to be able to produce a variety of parts. (Manager 3)

4	product mix changes		rescheduling capability

... if they change the plan, again, it comes to the flexibility and the people that we've got. We can fairly rapidly change over and reschedule to produce the parts that they need on time. (Manager 3)

Number of managers who mentioned the relationship	Uncertainty relating to...	⇒	Best way to cope with it is by developing...
2	parts and materials supply		materials control systems effectiveness

... anything that goes wrong with the schedule of material, straight away there is a warning bell from the system to our material control department and then they talk to the supplier... (Manager 1)

2	product mix changes		labour multi skills, transferability

...the broad range of skills our workers have gives us the flexibility to go and react to changes in the environment ...such as... program changes. (Manager 1)

2	overall volume changes		excess capacity

... we use some excess capacity to absorb some level of fluctuation in overall production volume. (Manager 4)

2	general uncertainty		ability to work in groups

The level of knowledge and experience out there has to be harnessed as a group, so I enjoy being supported by those groups and when they work well as a team, the problems I have I see resolved. (Manager 4)

2	general uncertainty		multi skills of labour, labour transferability
2	product introduction		integration design/ manufacturing

Number of managers who mentioned the relationship	Variety	⇒	Best way to cope with it is by developing...
2	variety		standardization of parts

Standardization is quite important to reduce variety and therefore the need to be flexible. (Manager 1)

2	variety		multi skills of labour, transferability

About the variety the flexibility I need comes from people. (Manager 4)

Some managers had a perception about the relationships between types of uncertainty and variability and ways to cope with them Yet, when asked about priorities, they did not seem to give priority to the actions they considered effective to cope with the uncertainty aspects they viewed as the most risky even when they had a low degree of certainty. There are two possible reasons for this behaviour:

i) they could be answering only what they perceived as expected answers about the relationships. This is not likely because of the measures the researcher took in order to increase the research instrument reliability; or,

ii) they may not have a consistent framework of reference to base their analysis on and therefore they cannot analyze the large amount of information they have in a systematic and comprehensive way. They would be able to perform analyses of parts of the overall problem but not to build a consistent decision model which takes into account the main variables and their inter relationships.

Manager 5, for instance, understands that parts supply and labour behaviour are the most risky uncertainty factors. He also considers these two as being among the most uncertain factors (he sees parts supply as only fairly predictable and labour behaviour as having low predictability). He also understands that rescheduling capability and development of multi-skills of labour are the best ways to cope with them respectively. Nevertheless, these two are not among his priorities or his critical success factors. He gave first priority to capability of machinery and excess capacity.

Another interesting point to notice is the emphasis given by the management in the achievement of manufacturing flexibility through the labour resource. At least three of the managers mentioned it:

> The main thing is - the flexibility is with the people. If people are not flexible, it doesn't matter how flexible your machines are, how flexible your processes are, if people aren't flexible then it's not worth anything, you remain static. (Manager 4)

Another point made by some managers, mainly the ones who were most concerned about flexibility, is that they see manufacturing flexibility as a sort of reserve, something that the organization possesses although it is not continuously using it. In the words of three of the managers:

> Flexibility is definitely an asset, something that is not currently being used but you can use when you need. I can use that asset, the flexibility to change things. It could be a reserve of ability, capacity or both. (Manager 6)

> Flexibility is like a commodity, something you have to possess, the willingness to change, the experience, the knowledge. ...It is a little accumulator of knowledge, abilities and capacity. It is an actual thing - either you have it or you don't. (Manager 2)

> Flexibility is like a reserve, a reserve that has been planned (Manager 3)

The relationship between the flexibility-related competitive criteria and resource characteristics or critical success factors In general, there seems to be consistency in the perception of the various managers interviewed in the flexibility-related competitive criteria they should pursue. All of them, for instance, ranked mix flexibility as the first flexibility-related priority. All but one of them specified mix response and one specified mix range as the particular mix flexibility dimension the engines plant should focus on. The consistency between the flexibility-related competitive priority and the resource characteristics which they considered critical success factors was also high. Figure A.3 shows the first two flexibility related priorities and the two first critical success factors according to the perception of each manager:

The only inconsistency seems to be with Manager 4 who places a high priority on machine capability; this can not be directly related to mix response or volume range, which are his priorities in terms of order winning. For the rest of the managers, the two columns seem to show consistency between system objectives and means to achieve them.

Table A.3 Relationship between flexibility-related competitive criteria and critical success factors, according to managers in Company A

Manager	Two first flexibility-related competitive criteria	Two first critical success factors
1	n/a	n/a
2	mix response delivery response	fast set-up times multi-skills, transferability
3	mix response volume range	change of line rates capability excess capacity
4	mix response volume range	capability of machinery excess capacity
5	mix response product response	re scheduling capability control systems effectiveness
6	mix range volume response	multi-skills, transferability ability to work in groups

Some conclusions of the within case A study

Managers in Company A see flexibility as a way to cope with uncertainties when the causes of such uncertainties cannot be eliminated. They seem to understand that variability and different types of uncertainty should be dealt with by developing different types of resource flexibilities. However, they do not seem to have a consistent decision model which includes different types of uncertainty, variability and different types of resource and system flexibility.

The most flexibility-conscious managers in Company A see flexibility as a reserve, something which has to be planned for, developed, maintained and seen as a valuable asset. In general, managers in Company A have a reasonable degree of consistency in their perception of the flexibility related order winning criteria they should pursue.

Case B - the Brazilian carburettor manufacturer (Company B)

Company B is a carburettor manufacturer located in São Paulo, Brazil. It is the main supplier of carburettors to the Brazilian car assembly companies and for the spare parts market. Company B is part of a large transnational corporation with headquarters in Europe and interests in a broad range of industrial products. Some figures about Company B are shown below:

Number of Employees	
Direct	788
Indirect	680
Number of product types	121 derivatives
approx % of components made in (in number)	40%
approx % of components bought in (in number)	60%

Organizational issues Company B plant is organized functionally, on a conventional job shop type of layout, although they are now at early stages of the migration into cellular manufacturing. Some pilot cells have just been established with promising results, according to the managers.

Company B is presently in a very particular situation. It has always been regarded as a carburettor manufacturer. Nevertheless, with the new (for the Brazilian market) technological advent of fuel injection, Company B has changed its mission into being an engine feeding systems manufacturer. The carburettor is going to die as an OEM (original equipment manufacturer) product in 1997, according to corporate plans. Therefore, no large investments are being made in the conventional carburettor technology and therefore no major changes in the line of carburettors are expected to be introduced in the future.

On the other hand investment is being made to qualify the company to compete in the new market, that of fuel injection systems. In order to do that, managerial and technical staff are being sent abroad, to be trained in the company's headquarters. Changing into the fuel injection technology is regarded by the managers in Company B as a major change. They reckon that the change is bound to bring many problems to the company since the new technology is based on microelectronics, rather than mechanics principles, thus demanding completely different skills, machinery and systems in order for the company to compete with other and comparatively more experienced competitors in the new market (The German Bosch, for instance). The change is supposed to be gradual, resulting in the end of the carburettor (except for the remaining spare parts market) in 1997.

Company B's organizational structure is conventional and hierarchical although they are currently trying to include some aspects of the matrix organization, establishing several multi disciplinary work groups with specific goals, such as product introduction, lead-time reduction, and so on, aiming at breaking the barriers between separate functions. There are statistical process control procedures implemented and the workers are responsible for the process quality, what they call self-control. Equipment maintenance is still performed by a separate team, though the very basic maintenance procedures are performed by the operators themselves. This arrangement also applies to the setting up of the machines.

The formal manufacturing planning and control system is basically MRP for the planning of materials and master scheduling. The scheduling however is done on a people-based informal system by the logistics manager and staff.

There are five directors under a general managing director in Company B: the industrial director (plant management, quality, maintenance, process development, production and after sales services), the materials director (supply management, production planning and control, materials handling and storing), the technology director (product research and development), the marketing director (relationship with the customers and market) and the finance and administration director (accounting, finance and sales). The human resources manager reports directly to the general managing director.

The general approach regarding industrial relations has recently changed, favouring the company. This change has partially been caused by the Brazilian recession, resulting in high levels of unemployment. The payment system for direct workers is based on grades, according to an internal merit assessment system, not directly linked to the breadth of skills of the worker nor to output rates.

The interviews Six people were interviewed in Company B:

The industrial director - referred to, in this case, as 'Manager 1', responsible for the plant management, maintenance, process development, production and quality assurance.

The logistics manager, referred to as 'Manager 2', responsible for the production and materials planning and control systems

The product engineering manager, referred to as 'Manager 3', responsible for the product engineering

The production manager, referred to as 'Manager 4', responsible for the production

The industrial technology manager, referred to as 'Manager 5', responsible for process design and development

The quality control manager, referred to as 'Manager 6', responsible for the quality control and assurance engineering

Line of products: variety, variation and innovation Company B currently has a line of seven basic product families, with minor to considerable differences between products within a family, depending on the specific family. The overall number of products or 'derivatives' is 121. The variation in overall volume can be approximately 50 per cent from month to month. The variation in the mix of products demanded can be very high. For instance, in each week 30 out of the 121 derivatives are produced. Moreover, every month at least 60 per cent (or approximately 72) of the product range is produced.

The introduction of new products or the engineering changes of the existing ones are done on a continuous rather than on a batch basis. Six to eight engineering changes are made each month, 50 per cent being minor changes concerning process improvement and 50 per cent changes in the application of the products, owing to changes in the fuel composition, emission regulations and other customer requests.

Launching a completely new carburettor is not in the future plans of the company, since the carburettors have a certain date to die as an OEM product in Company B. Although the managers consider that Company B's performance in the past, in terms of introduction of new products and product changes, has been clearly better than that of the competitors, this has been accomplished at very high costs in terms of resources and organizational disruption. In the words of the industrial director:

> It is somewhat similar to a football match played by kids. The whole team is always running after the ball, disregard of their positions; when something crops up, everybody suddenly change priorities and start running to try to 'fight the fire'. It looks like flexibility, but it is not, because the effort it is poorly coordinated, not planned for and very stressful. (Manager 1)

Historically, the time period to introduce a completely new product has been two years. Nevertheless, as a preparation for the new line of products, the fuel injection systems, the company has established a task force to develop a new system for the introduction of new products ('sistema Company B de novos produtos', or 'Company B system for new products'), trying to incorporate concepts of multi-functional work groups and simultaneous engineering (which are relatively new concepts in Brazil). The system now exists in the form of a written document but most of the managers recognize that there is still a long way to go in terms of breaking the barriers between functions and making it work fully. Managers consider that the ability to introduce new products quickly and reliably (in terms of quality) will play a major role in the future competitive scenario.

Manufacturing flexibility task and performance The managers' answers concerning the flexibility-related task of the plant are consistent, to a certain extent. Four out of six managers specified product range as the particular flexibility dimension the company should primarily focus on. The other two managers (Managers 4 and 6) mentioned delivery range and mix response as their first competitive priorities respectively. Table B.1 shows the distribution of answers. The figures under the 'Mgr' columns represent the priority each of them gave to the individual flexibility types in terms of order-winning criteria. One of the managers (Manager 2) did not feel comfortable in ranking the criteria. Rather, he preferred only to mention the ones he considered relevant. That is the reason why in his column there are four numbers '4'.

Table B.1 Priority given by managers of Company B regarding types and dimensions of manufacturing flexibility

	Mgr 1	Mgr 2	Mgr 3	Mgr 4	Mgr 5	Mgr 6	Total	Total weighed
Droduct range	4	4	4		4		4	16
Product response	3	4	3				3	10
Mix range		4					1	4
Mix response						4	1	4
Volume range					3		1	3
Volume response						3	1	3
Delivery range		4		4			2	8
Delivery response				3			1	3

Uncertainties involved The uncertainties mentioned by the managers as the ones which represent the highest potential risk to Company B's competitiveness show a distinct pattern. All the managers, for instance, placed uncertainties regarding materials and parts supply as one of their two main concerns in Company B. Four of the managers placed management behaviour under changing circumstances among their two main concerns. Other aspects mentioned among the two main concerns are uncertainties with regard to machine breakdowns, to specification of new products and to the availability of technological information. The perceived uncertainties in Company B can be found in Table B.2.

Table B.2 Importance given by managers of Company B regarding sources of uncertainty as risky for competitiveness

	Mgr 1	Mgr 2	Mgr 3	Mgr 4	Mgr 5	Mgr 6	Total	Total weighed
parts supply	4	4	4	4		4	5	20
managers behaviour	3	2		3	3	3	5	14
machine breakdown	1	3		2		1	4	6
quality of products	1	1		1		1	4	4
products introduction		1	3				2	4
mix changes		1		1	1	1	4	4
volume changes					1		1	1
technologic. resources					4		1	4

It seems that the uncertainty sources which concern the managers most are those related first with parts and materials supply and second, interestingly, with the management behaviour under changing circumstances, here understood as the unpredictable response by supervisors and middle managers to changes in current practices. To a certain extent this assessment can be explained by the difficulties the managers predict they will have with the major changes the company is going to face in the near future, with the introduction of the fuel injection systems technology.

Coping with change, uncertainties and variabilities When asked how they coped with the different types of uncertainty and variability, the managers showed, in general, both different approaches and different levels of understanding of the variables involved with manufacturing flexibility. The managers in Company B also seemed to show a greater concern with issues related to product quality than with flexibility.

The most frequently mentioned relationships between types of uncertainty, variability and ways to deal with them are shown below, together with some representative quotations regarding them. The first column represents the number of managers who mentioned the relationship. The second and third columns represent the relationship itself, the second being the source of the uncertainty and the third, the capabilities or ways managers see as worth developing for coping with the uncertainty shown in the second column.

Number of managers who mentioned the relationship	Uncertainty relating to...	⇒	Best way to cope with it is by developing...
4	parts and materials supply		rescheduling capability

...we try several alternatives, we analyze the impact of the delay and, if that is the case, we re-schedule and do whatever product we can ... (Manager 4)

4	parts and materials supply		supplier development, partnership

...we have some plans to overcome these difficulties [of having an uncertain supply] and they consider the supplier as a partner; we have to work together with them, we have to pass on the idea that if Company B is successful, the supplies will also profit...so, in the long term, the idea is to reduce the supplier base and develop co-operation rather than confrontation... (Manager 4)

4	management behaviour		training, awareness

...the middle management is considerably more resistant to change than the direct labour. If a new idea is proposed, in maybe 50 per cent of the times, the middle managers react against it, sometimes with no apparent reason. The way to deal with this is by training them, increasing the level of their awareness, and, as a last resource, substituting them... (Manager 1)

3	labour behaviour		multi-skills development

...to prevent against lack of continuity, caused by absenteeism, we have to develop a multi-skilled workforce... (Manager 1)

3	product mix changes		set-up times reduction

...because the mix changes much and frequently, with short set-up times it is much easier to respond... (Manager 6)

3	quality changes		interface prod design/ proc design/manufact

Number of managers who mentioned the relationship	Uncertainty relating to...	⇒	Best way to cope with it is by developing...

...we have to have a good system to guarantee the integration of these functions because shortening the time for designing gives the process designers time to conceive the process properly; otherwise, the scrap rates will be higher and the probability of achieving good conformance will be lower... (Manager 6)

2	product mix changes		rescheduling capability

...we have to be good at rescheduling to fight the instability of the demand mix (Manager 5)

2	machine breakdown		re-routing capability

...to face it I have to have alternative machines, alternative process routings... (Manager 2)

2	product introduction		integration design/ manufacturing

...we have to invest in design technology to be able to deal with changes which are requested at the last moment, so you have to be very quick in designing and therefore you give more time and alternatives for the process designers and production people... (Manager 3)

2	variety		standardization of parts
2	product mix changes		ability to get/ maintain lead times low

...[we need short supply lead times] to keep our own lead times short. With shorter lead times your programming is more flexible, you can change the program with short notice and respond better to changes in the customer demand. (Manager 4)

2	parts and materials supply		machine capability

Number of managers who mentioned the relationship	Uncertainty relating to...	⇒	Best way to cope with it is by developing...

...when a part arrives and it is below the quality specified, we assign one person to try to find alternative solutions...can we use the part for that particular case? can we correct the problem by reworking the part using our internal capability? - and that is very frequent... (Manager 6)

2	labour behaviour (absenteeism)		extra capacity

...we program a load which is lower than the total system's capacity, because of the uncertainty with the absentees... (Manager 2)

2	machine breakdowns		ability to organize over time/subcontract

...besides alternative machines, I have to have the alternative to outsource the part quickly... (Manager 2)

2	materials and parts supply		ability to organize over time/subcontract

...sometimes I can rework the part in, but because I hadn't programmed it, I have to do it in overtime...(Manager 1)

2	quality		ability to work in groups

One point which is worth mentioning is the lack of emphasis given by the managers on the achievement of manufacturing flexibility through the human resource. Technological (fast set-ups, capability of machinery) and infrastructural resources (integration design/production via a formal system, re-routing the production flow ability through a system) seem to play the major roles in the view of Company B's management in achieving system flexibility.

Another point is the great emphasis the majority of the managers place on quality. Some unexpected associations of the resource aspects (meant to represent characteristics associated with flexibility) with quality appeared. An example of this is the importance of having an agile product design function in order to give the process design function time to design a proper process which can guarantee the product quality, rather than the more obvious relationship that fast design keeps with fast product introduction (Managers 2, 3 and 6).

That association may be caused by the present stage in which the company finds itself, still struggling with quality problems. If this is the case, De Meyer's (1986) hypothesis that there are stages which the companies progressively go through is confirmed. De Meyer found out that flexibility had not yet become a major competitive priority for the American manufacturers of his sample, whereas it was a major priority for the Japanese companies. De Meyer suggests that the reason could be that the American managers would be subscribing to the view that a flexible response to competitive threats is only possible if the basic quality and process problems are solved. Company B would still be struggling with basic and process problems and therefore flexibility would not yet be one of their priority concerns.

Furthermore, it appears that some of the managers see flexibility as something where they are 'forced' to possess to cope with uncertainties. Ideally, they would prefer to control the causes of the uncertainties, aiming at reducing them. However, since this is not always easily achieved in the short term and also since it is impossible to eliminate completely the stochastic components of the processes, they are 'forced' to develop flexibility.

Four of the managers (Managers 1, 2, 3 and 4) pointed to the need to co-operate with the suppliers in the long term (or in other words, to increase the coordination between Company B and the suppliers), in order to reduce the uncertainties that Company B has to deal with. In this case, they see the flexibility of the process as a means to deal with the effects of such uncertainties in the short term.

The same happened to the uncertainty regarding machine breakdowns. Three managers (Managers 1, 4 and 5) mentioned that in the long term preventive maintenance should be used to reduce the uncertainty level of the process continuity. Again, in the short term they point out flexibility-related solutions such as alternative routes and multi-capable machines to deal with the effects of the uncertain events.

The relationship between flexibility-related competitive criteria and resource characteristics or critical success factors In general, there seems to be some consistency in the perception of the various managers interviewed about the flexibility-related competitive criteria they should pursue.

Four of them (Managers 1, 2, 3 and 5), for instance, ranked product flexibility as the first flexibility related priority Company B should pursue. All of them specified product range as the particular product flexibility dimension the plant should focus on. The other two (Managers 6 and 4) ranked mix and delivery flexibility as the priorities. The consistency between the flexibility-related order winning criteria and the resource characteristics which they considered critical success factors was also high. Table B.3 shows the two first flexibility-related priorities and the two first critical success factors according to the perception of each manager.

Table B.3 Relationship between flexibility-related competitive criteria and critical success factors, according to managers in Company B

Managers	Two first flexibility-related competitive criteria	Two first critical success factors
1	product range product response	integration design/production ability to work in groups
2	product range product response mix range delivery range	rescheduling capability integration design/production
3	product range product response	integration design/production supplier development
4	delivery range delivery response	fast setups ability to work in groups
5	product range volume range	standardization rescheduling
6	mix response volume response	integration design/production fast set-ups

The only apparent inconsistency regards the answers of Manager 5 who places a high priority on rescheduling capability and standardization of parts which can not be directly related to product range or volume range which are his priorities in terms of competitive criteria.

For the rest of the managers, the two columns seem to show consistency between system objectives and means to achieve them.

Some managers had a definite perception about the relationships between types of uncertainty and variability and ways to cope with them. Yet, when asked about priorities, they did not seem to give priorities to the actions they considered effective for coping with the uncertainty aspects they viewed as most risky even when they were considered uncertain. Two among the possible reasons for this behaviour are:

i) they may not have a consistent framework to base their analysis on and therefore they cannot analyze the large amount of information they have in a systematic way. They would be able to perform analyses of parts of the overall problem but are not able to build a consistent decision model which takes into account the main variables and their inter relationships; or,

ii) they may give more importance to aspects related to the aggressive response to the market needs than to the preventive development of 'safeguards' against risks to the company's competitiveness. This can be noticed by the high consistency between their ranking regarding types of system flexibility and types of resource flexibility. The managers can identify factors which represent risk to the company's competitiveness. They also have an idea about which 'antidotes' would deal with them, but they do not give the same strategic importance to these 'antidotes' (or in other words what we could call defensive competitiveness) as they give to the factors which lead to what they perceive as market needs. It would be somewhat similar to a coach who emphasizes the development of the team's attack, not giving the same emphasis to reinforcing the team's defence.

Manager 1, for instance, understands that parts and materials supply and management behaviour under changing circumstances are the most risky uncertainty factors. He also places these two factors among the most uncertain ones (he perceives parts and materials supply as being only fairly predictable and management behaviour as having low predictability). He also understands that a number of ways can be used to cope with such factors. (He coherently regards supplier development, ability to reschedule, some excess capability and capacity as appropriate for coping with uncertainties relating with supply and training as a way to cope with uncertainty in the behaviour of the management.) Nevertheless, none of these factors are among his priorities or his main critical success factors list. He gave first priority to integration product design/process design/production and to the ability to work in groups, both not directly linked to his 'remedies', although highly consistent with the way he sees how the company competes in terms of flexibility (his first and second priorities are, respectively, product range and product response).

Such a pattern of perception is somewhat general among the managers. Summarizing, it seems that managers see the relationships between flexibility-related order winning criteria and the critical success factors to achieve high performance in them. They are apparently able to focus their attention and give priority attention to these factors. Nevertheless, less importance seems to be given to the factors which would represent the insurance or the safeguards against the uncertainty factors considered by themselves as risky to the company's competitiveness.

Some conclusions of the within case B study

Managers in Company B see flexibility as a way to cope with uncertainties when the causes of such uncertainties cannot be eliminated. It seems to be clear for them that different types of uncertainty should be dealt with by developing different types of resource flexibility. They also seem to have a greater concern towards quality issues than towards flexibility issues.

Managers in Company B seem to place more importance on the 'competitive weapon' aspect of flexibility, something which could be aggressively and proactively developed and sold, than on the safeguard against the uncertainties aspect. Managers' perceptions in Company B have a reasonable consistency in terms of the flexibility related competeitive criteria which they should pursue.

Case C - the Brazilian shock absorber manufacturer (Company C)

Company C manufactures and distributes, to the automotive market, components having a high technological content. It is an entirely Brazilian-owned company whose capital is open to the general public and whose shares are traded on the country's stock exchanges. As the largest domestic producer of automotive parts, it ranks 71st, based on sales, among private sector companies in Brazil.

Company C aims at the high technological content automotive parts market and with this objective invests approximately 3 per cent of its operational revenue in product and process research and development. Some figures about Company C's Group referring to 1988 are given below.

overall turnover	US$ 500 million
market breakdown-invoicing	
original equipment	42%
spare parts market	40%
exports	18%
market share in Brazil	
piston rings	92%
shock absorbers	75%
castings	60%
employees	15 thousand

Organizational issues Company C is organized in divisions. There are six main industrial divisions: shock absorbers, engine components, castings, exhaustion systems (mufflers), sintered parts and polyurethanes.

In 1987, Company C began operation of its first production cells in various divisions, as part of a comprehensive group program called 'programa de qualidade total Company C', or 'Company C total quality program'. That includes statistical process control implementation, cell manufacturing, set-up reduction programs, MRPII implementation and better industrial relations. To support the program, an ambitious training program was designed in which more than 10 thousand men/days per year are dedicated to off-the-job training.

The results so far have been considered satisfactory by the managers. They now have a number of cells in operation. They also claim reductions in the average production lead time for piston rings, for instance, from 25 to 14 days, 5 per cent reduction in work in progress and substantial (not quantified) improvement in conformance quality levels.

The operators are nowadays in charge of the cleaning of the work place, basic machine maintenance and statistical process control.

The formal manufacturing planning and control system is MRP II for the planning and control of materials supply and inter-cells coordination. The dispatching and very short term shop floor control activities are done by special task forces, called 'follow-up teams', responsible for keeping up with recent program changes.

The industrial relations and payment schemes are conventional, the payment of direct labour is linked to good output levels and the approach to benefits is considered by one of the managers as *patronizing*. According to him, this is a trade mark of the group's founder-president. Emphasis is given to training, but not to multiskills development, which is to a certain extent unusual when companies migrate to cell manufacturing. The relationship with the powerful 'ABC'[2] Unions has not been very smooth, with a number of disruptive strikes cropping up during the last years.

The divisions are reasonably autonomous, with division directors leading teams of within-division managers. Nevertheless, when the group decided to implement the reforms in production processes, a new post was created, that of director of productivity and quality (who is one of the interviewees in this case-study), who reports directly to the group's president. The recently appointed director assembled, then, a multi-functional, multidivision team to be the agents of change within each division, aiming at implementing the planned manufacturing changes.

The interviews Three people were interviewed in Company C:

The operations manager of the shock absorber division, responsible for the division's materials flow, industrial engineering and quality assurance. In this case-study, he is referred to as 'Manager 1'.

The senior sales manager of the shock absorber division, responsible for the division's relationship with the market and interface with the operations function, which is very intense in Company C. He has been working for the Company C group for a number of years and has actually worked for most of the divisions in various positions including production management. he has broad experience and knowledge of the group. Here he is referred to as 'Manager 2'

The director of quality and productivity of the group, responsible for the design and implementation of the Company C total quality program, an ambitious program aimed at making Company C a world class manufacturer. The shock absorber division is leading the quality program and has served as a 'lab' for pilot studies for the group, in terms of new techniques. He is 'Manager 3'.

The case-study will focus on the shock absorber division, although some of the examples given by the interviewees refer to facts which happened in other divisions.

Line of products - variety, variation and innovation The shock absorber division has approximately 2 thousand active products, according to the operations manager, and they are in general similar products which are also not very different from each other in terms of process.

A shock absorber has about 30 different parts and components. In Company C more than 90 per cent of them are made in. The company used to buy some components, mainly sintered and polyurethane parts. Nevertheless, as part of the group's policy of vertical integration (another 'trade mark' of the president, according to one of the managers) to reduce transaction costs, the group bought out two companies in the 1980's which are today the sintered parts division and the polyurethane division.

Approximately 150 product changes are performed each year in the shock absorber division, 20 to 30 being new designs. The changes are not made in batches but on a continuous basis.

The process of launching a new product, at the time of the case-study interviews, was done in a conventional fashion, with well defined sequential stages of product design, process design, prototyping and finally production. CAD is in the company's plans but it has not been implemented yet.

The variation in volume is what seems to worry the operations manager most, since the group have an aggressive policy aiming at new export markets. The overall demand can vary 30 per cent.

> Last week, for instance, an American buyer came to us and ordered 128 thousand shock absorbers. That is 10% of our annual production. Now I have to decide what orders I will delay, because we are bottlenecked, working in three shifts. We will have to struggle to deliver in the 4 months period we promised... (Manager 1)

That is a concern, especially because the factory is *bottlenecked*, with very high occupation rates. This is due to another policy of the group, which is the *chasing-the-demand* policy for investments. Investments in new equipment are made only when there is a guarantee that the equipment will be fully utilized. The group can afford to do that, mainly in terms of the domestic market niche where Company C operates, which is a seller's market. The company is virtually a monopolist in one of the product lines (piston rings) and almost so in others (75 per cent of the shock absorber domestic OEM market and 85 per cent of the spare parts market, for instance).

Manufacturing flexibility task and performance Regarding the manufacturing task in general, some inconsistencies were noticed between two managers' (Manager 1 and Manager 2) views, which reflects a possible communication problem and/or lack of a uniform understanding of the company's manufacturing task. In the words of the two managers:

> For the export market, the order winning criteria is price. Nothing else. Quality is needed anyway. You must have high quality just to qualify for the market. Delivery speed is not relevant either... (Manager 1)
>
> The export market represents 30 to 40 per cent of our turnover, we are competing at the world level. At this level, it is not enough that the products have quality and price. They have to be delivered fast and on the right time. I am in contact with this market every day, and they want fast and reliable delivery. I have just had a meeting with an Italian customer who came to complain about our delay in performing a modification he ordered... We have to improve our times... (Manager 2)

The two interviews above were done in the same week. Regarding the flexibility-related manufacturing task, two out of three managers specified new product flexibility as the particular flexibility type the company should primarily focus on (response followed by range). The other manager (Manager 3) mentioned delivery response and mix response as his first and second competitive priorities, respectively. This inconsistency may be caused by a lack of common understanding of the company's manufacturing strategic task. Since all of them gave great importance to fast response to customer orders, the lack of agreement seems to be between serving the orders for existing products better or giving priority to winning new orders for new products. Table C.1 shows the distribution of answers.

Uncertainties involved The uncertainties mentioned by the managers as the ones which represent the highest risk to Company C's competitiveness are basically related to two aspects: the supply chain and government intervention. According to the managers, the lack of clear and stable rules and policies, set by the government, and under which the company has to operate, affects several aspects of the company's operations, such as export market demand. They argue that, because of lack of consistency between the progression of inflation rates and exchange rate mechanisms, for example, sometimes what seems to be good and profitable business, at the time a deal is set up to export goods, becomes a loss at the time you actually deliver the goods and receive the payment. In the words of Manager 3:

> We never know what will be our revenue with exports and our expenses with imports. It is tough planning in such an environment... (Manager 3)

That gives product prices an uncertain component and sometimes causes a sudden increase in demand because the company suddenly becomes more cost-competitive in external markets due to an unexpected change in the exchange rates.

Table C.1 Priority given by managers of Company C regarding types and dimensions of manufacturing flexibility

	Mgr1	Mgr2	Mgr3	Total	Total weighed
product range	4	4		2	8
product response	3	3	1	3	7
mix range				0	0
mix response		2	3	2	5
volume range				0	0
volume response		1	2	2	3
delivery range				0	0
delivery response			4	1	4

Managers 2 and 3 ranked government intervention as the most risky uncertainty source for Company C's competitiveness. Manager 3 ranked 'demand uncertainty' as one of his main concerns, which in a way is related to the government intervention aspect, as explained earlier. He also considered uncertainty with 'parts and materials supply' in terms of delivery times as risky. However, although the operations manager mentioned the problem as a local one ('in general, 10 to 20 per cent of the supplies miss their delivery dates'), it could well be the case that the problem is not with the suppliers alone. As Manager 3 puts it:

> About the suppliers, we don't have major problems in terms of quality. Nevertheless, we demand more flexibility from them (in terms of volume and delivery) than they can cope with. It is difficult therefore to identify who is responsible for the faulty deliveries. The uncertainty of the demand end of the chain, which is linked to the uncertainties with the government policies, is transmitted backwards and destabilizes the supply side of the chain. I'll give you an example: we developed a new product for a new car. The initial demand forecast of our customer was 5 thousand products per month. We contacted our suppliers and they quoted the raw material and parts we would need, and they got prepared for a demand of 15 thousand products per month. That is because our material planners knew that we weren't the exclusive suppliers of our customer and we know that historically we end up winning more share making our demand go up. So they had planned 15 thousand. Well, two months ago the customer called us to complain that his assembly line had to stop because they were short of the product. I went to check how many products they were consuming:

48 thousand products per month. And that was after five months. How can I complain with my suppliers? And I can't complain with the customer either. He is struggling with his own problems. We have to do our best to accommodate the situation ... (Manager 3)

The managers also mentioned several other uncertainty sources, but they were quite clear that their first and second ones (actually only the first one in the case of Manager 2, who regards all the others as mere consequences) were the most relevant.

The uncertainties perceived as the most risky by the interviewed managers in Company C can be found in Table C.2.

Table C.2 Importance given by managers of Company C regarding sources of uncertainty which represent risk for the company's competitiveness

	Mgr1	Mgr2	Mgr3	Total	Total weighed
parts supply	1	4		2	5
labour behaviour/supply	1	2		2	3
equipment supply	1			1	1
machine breakdown	1			1	1
technological information supply			1	1	1
product introduction	1			1	1
mix changes	1		3	2	4
volume changes	3		1	2	4
information systems			1	1	1
government intervention		4	4	2	8

Coping with change, uncertainties and variabilities The most frequently mentioned relationships between types of uncertainty, variability and ways to deal with them are shown below, together with some representative quotations regarding them. The first column represents the number of managers who mentioned the relationship. The second and third columns represent the relationship itself, the second being the source of the uncertainty and the third, the capabilities or ways managers see as worth developing to deal with the uncertainty shown in the second column.

Number of managers who mentioned the relationship	Uncertainty relating to...	⇒	Best way to cope with it is by developing...
2	parts and materials supply		improve coordination with suppliers

Today, we use buffer stocks to prevent running out of critical parts, but with MRP we intend to improve our coordination with the suppliers, we will give them a delivery program ... (Manager 1)

We have established longer term contracts and now we are working together with them (the suppliers) to develop ways of improving the dependability of their delivery. (Manager 2)

2	labour supply		training

...because we have adopted more sophisticated production processes, we can no longer find as many people as we need ready, in the community. As a consequence we had to intensify our training programs. So we recruit people without qualifications and make them qualified workers by training them within the company. We have recently bought a school in which we subsidize the education of 4 thousand children and teenagers. We teach them professional skills at the levels we need ... (Manager 2)

2	machine downtime		preventive maintenance

...we have an occupation rate which is too high and we don't have as much time as we needed to do preventive maintenance, which is preferable. Then we have to work fast to identify the problem and put the machine up again (Manager 1)

2	government intervention		short lead times

...if I have long cycle times, I make myself more susceptible to the effects of uncertainties, such as the exchange rate mechanisms. The strategic issue nowadays is to be fast. So we have to work on the time elapsed since the customer orders a change in one of its products until the production acknowledges it, we have to work on the time that it takes for us to reschedule and accommodate a change in demand the time our supplier takes to react to the changes we request and so on. We have to reduce all the cycle times involved in our operation. The nervous system in this regard is the flow of information. We have to create information systems, and I am not talking about large information systems which know everything and decide everything. The idea that you perform well with an intelligent mind and an obedient body is not valid any more. Now each part of the body has to be able to react as well as proact. (Manager 3)

Number of managers who mentioned the relationship	Uncertainty relating to...	⇒	Best way to cope with it is by developing...
2	volume changes		forecasting systems

I have done an analysis of our forecast deviations and I concluded that the main factor which caused deviation in our forecasts is the circumstance in which the forecast was made. If the forecast was made in a good period, it was in general optimistically biased and the other way round. They were too nervous. Good volume demand forecasting is very important for us... (Manager 3)

2	unions behaviour	monitoring environment

...we have to be constantly monitoring the situation with the Unions, otherwise, we are taken by surprise by a strike... (Manager 3)

2	new product introduction	ability to subcontract

We have this culture of vertical integration, of doing everything in house and that constrains us. If we were good at subcontracting, we could respond quicker to customer needs. We now have a queue of six months in our die machining shop. (Manager 2)

A number of other relationships were mentioned by not more than one manager. Nevertheless, they are worth mentioning:

1	information flow	buffer stocks

Today materials planning and production are not well coordinated. Because of that, we keep some stock buffers to react better. (Manager 1)

1	demand mix	reduce set-up times

> We now have a large set-up reduction program running. The goal is to reach two minutes for every set-up. When we get there, we will be able to respond to changes in demand mix... (Manager 1)

When asked how they coped with the different types of uncertainty and variability, the managers showed, in general, a similar approach. Perhaps because of the company's conservative culture, mentioned by all the managers interviewed, they put a greater emphasis on developing ways to control the sources of uncertainty than on developing ways to deal with the effects of such uncertainty.

Among the ways used by the Company C managers to control the environmental change, vertical integration seems to play an important role. They vertically integrated the supply of parts (sintered and polyurethane parts), equipment (a division was established to manufacture machines for the group to avoid the usual problems with supply), labour (they acquired a school to educate people from the community at the levels they needed) and technology (Company C has established a large research and development centre to reduce the need to rely on other companies as their technology suppliers).

Concerning the relationship with their suppliers, after vertically integrating up to the point at which almost 100 per cent of the parts were made in, they still try to control further the uncertainties with suppliers by improving the coordination between Company C and its suppliers, by means of establishing long term contracts and co-operation. In terms of demand improving forecast systems is considered another important way of controlling the predictability of the changes the company has to deal with.

It seems that, generally, the managers see flexibility as something they have to develop to cope with the uncertainties or, more broadly, the changes which have not been reduced or eliminated. Ideally, though, they would rather control the causes of the uncertainties by aiming at reducing them, but since this is not always easily achieved in the short term they are 'forced' to develop flexibility.

Managers in Company C seem to rely more on infrastructural resources (particularly systems) than in human or technological resources to achieve desired levels of system flexibility. The three interviewed managers mentioned the need to develop more responsive systems, whereas they considered that labour multiskills are not very important for Company C's operation. Manager 3 put it this way:

> Resistance to change is the normal attitude amongst human beings. The companies which respond better to change invariably are the ones which created systems to be more flexible. When you change, you always create problems. To change, people have to be convinced that the pain for

remaining unchanged is worse than the pain of changing... so, the only way out is to install a system designed to be dynamic, which guarantees the change. If companies could live in an 'aseptic bubble' free from environmental stimuli, there wouldn't have to be change. The company changes when the environment in which it is immersed 'pushes' it. From then on, it is a matter of managerial competence to drive the change... (Manager 3)

The relationship between flexibility-related competitive criteria and manufacturing-related critical success factors There seem to be some inconsistencies in the perception of the various managers interviewed about the flexibility-related competitive criteria. Two of the managers (Managers 1 and 2) ranked product response flexibility as the first flexibility-related competitive priority Company C should pursue, seconded by product range flexibility. The third manager (Manager 3) ranked delivery response and mix response as the ones he considers to be priorities.

The relationship between the flexibility-related order winning criteria and the resource characteristics which they considered to be critical success factors is listed below. Table C.3 shows the two first flexibility related priorities and the two first critical success factors according to the perception of each manager.

The two columns for Managers 2 and 3 do not seem to show clear inconsistencies between system priority objectives and factors they considered critical for the company's competitive success.

Manager 1, however, saw new product flexibility as the main flexibility-related competitive criteria but his critical success factors do not seem to bear a close relation with them.

Table C.3 Relationship between flexibility-related competitive criteria and critical success factors, according to managers in Company C.

Manager	Two first flexibility-related competitive criteria	Two first critical success factors
1	product response	fast set-ups
	product range	re scheduling capability
2	product response	fast design
	product range	standardization
3	delivery response	information flow
	mix response	coordination with suppliers

They are related, though, to the uncertainty factors which he pointed out as the most risky ones (materials and parts supply and demand variation). This could well indicate that Manager 1 has priorities which primarily aim at reducing the risks for the company's competitiveness. Managers 2 and 3, on the other hand would have priorities which aim at winning orders in the market place (priorities which are consistent with their flexibility-related competitive criteria). It could also mean that not all the managers have a consistent framework to help them establish priorities which are consistent with the overall company objectives.

Some conclusions of the within case C study

Managers in Company C see flexibility as a way to cope with changes when the causes of such changes cannot be eliminated. They understand that different types of changes should be dealt with by developing different types of control systems and/or resource flexibilities

Managers in Company C do not seem to have a consistent decision model which includes different types of uncertainty, variability and different types of resource and system flexibility. Their perceptions have a reasonable consistency in terms of the flexibility related order winning criteria they should pursue and the ways they should achieve them. Nevertheless, there are discrepancies regarding which are the flexibility-related competitive criteria for the division. Managers in Company C do not seem to have a clear view about the differences between controlling the uncertainties and dealing with the uncertainties.

Case D - the English vehicle manufacturing plant (Company D)

Company D is a vehicle manufacturing plant located in the Midlands, England, and it is part of a large transnational corporation with head-quarters in North America and interests focused on automotive products, industrial machinery and engines. It is one of the largest factories in the world dedicated to the production of that class of motor vehicle and it specializes in the design, manufacture and supply for worldwide markets. 90 per cent of the 65 thousand vehicle sets produced at Company D's plant each year are exported to over 140 countries. The annual turnover of the plant is approximately 120 million pounds. Some figures with regard to Company D are shown below:

Number of Employees	
Direct	1600
Indirect	400
Number of possible vehicle configurations	3640
approx % of components made in (in number)	15%
approx % of components bought in (in number)	85%

Organizational issues The Company D plant is organized functionally, on a job shop type of layout, although they are now at early stages of the migration into cellular manufacturing. Two pilot cells have just been established with promising results, according to the managers.

Company D's organizational structure is hierarchical although they have recently gone through organizational changes. Such changes included the substitution of a number of directors, the re-design of the organizational chart, the inclusion of many aspects of the matrix organization, with the establishment of several multi-functional groups with specific goals, aiming at breaking the barriers between separate functions. Presently, the team members are dedicated to the projects on a part time basis, keeping links with their functional departments. According to the managers, the dedication of the members to the projects is planned to become full time in five years.

There are statistical process control procedures implemented and the workers are responsible for the process quality. Equipment maintenance is still performed by a separate team, although the very basic maintenance procedures are performed by the operators themselves. This arrangement also applies to the setting up of the machines.

The formal manufacturing planning and control system is MRPII, although the managers consider the use of MRPII to be an intermediate stage towards the JIT production. The day-to-day changes in the schedules nevertheless are done by an informal system, because, according to the managers, the MRPII software packages do not provide the company with the flexibility it needs to cope with its broad product range and highly variable demand.

There are five directors under a general managing director: the manufacturing director (plant management, quality, maintenance, process development, production and after sales services), the director of supply (supply management, production planning and control, materials handling and storing), the technology director (product research and development), the marketing director (relationship with the customers and market) and the finance and administration director (accounting, finance and sales).

The interviews Four people were interviewed in Company D:

The supply director, referred to in this case as 'Manager 1', responsible for the management of the supply system, relationship with suppliers and related issues.

The product design manager, 'Manager 2', responsible for the product design and development and management of the bills of material.

The advanced manufacturing engineering manager, 'Manager 3', responsible for the analysis and design of the manufacturing systems

The production manager, 'Manager 4', responsible for the production and plant management

Line of products - variety, variation and innovation Company D builds vehicles to order and currently has a line of two basic product families, with considerable differences between products or configurations within a family. The overall number of products or derivatives is theoretically 3640, of which approximately 2 thousand are made in any one year, considered exaggerated by two of the managers interviewed:

> I think we offer too many product variables, but they say that is what the market wants... (Manager 3)

A third manager (Manager 1), on the other hand considers product variety as the main competitive advantage of Company D. The variation in overall volume can be approximately 20 per cent from month to month. The demand is seasonal and the variation in the mix of products demanded is also high.

The introduction of new products or engineering changes of the existing ones is done on a continuous rather than on a batch basis. In average, thirty minor engineering changes are made each month and one substantial change per quarter in functional aspects of the products.

Historically, the time period for introducing a totally new product has been five years. Nevertheless, according to the managers, the company has recently made efforts in creating the conditions for the simultaneous development of new products, with multi-functional teams participating in the process since the early conceptual stages, to ensure 'design for manufacturing'. No results have yet been noticed in terms of time to develop a product, partially because the emphasis has been on designing out unnecessary variety. Even so, the managers believe that reductions in time will soon follow. Some managers commented on the issue of launching new products and also on the problems they are finding with implementing simultaneous development concepts:

> If you get the right type of people and put them together at the right time to design the products, you reduce the causes of complexity and the variability, you design out long lead-times, design out complexity, variability and this way you solve a lot of the problems at the back of the system. (Manager 3)

> Company D decided to put a tremendous amount of effort in simultaneous engineering because we believe that we can get the turbulence down, 30 to 40 per cent and still keep the product variety. The effort is to make the products to vary only in the bits which the customers need to be different. The main aim of the program is to make the manufacturing more effective and responsive ... controlling factors which are not purely design factors such as the bill of materials, not exploding the bill of materials, not exploding the variability of parts... We do it by creating an environment without any brick walls. We have to build a team, a disciplined team, with people from different parts of the business all brought in at the conceptual stage... All the concepts have to be addressed: reducing costs, shortening lead-times, design, production, supply, etc... We actually work physically together. The leadership has to be from the top. The directors of the company and the managing director, they've got to say: 'I want to do it', they've got to allow people time, etc. People are part-time in the teams. We needed a cultural change because a new perspective is needed. (Manager 3)

> The general process (of product introduction) is that there is an engineering proposal responding to a marketing request. When the proposal has been bought off, we have some economics done on it. These analysis have been done for the last couple of years by project teams, across functions. They have been very good at getting to launch the product up to the prototyping but useless to getting the thing into production volumes. As soon as we get the project 'go-ahead', because it makes economic sense, engineering will start to go into detailed design and development. That will continue to be reviewed by marketing, purchasing and manufacturing. But because generally they are too busy with day-to-day activities, they don't want to get too involved at this stage. They don't have the design engineer mentality, they want absolute detail, definition before they say: yes, I am

happy or not. But when they are not, it is often very late in the process. Some of it is caused by lack of understanding of the process and some of it is caused by the way things have always been done. This department used to be five miles down the road. There was a large gap. Actually everything used to be designed in America and that was absolutely awful. Besides, the person who comes to take part in the project team can be a problem because in general he is not the representative of his department, but of the small part of the department he works in. We are not very good at freezing the design. We are also not very good at launching new products... Then we make the formal release to manufacturing, the bill of materials, then manufacturing do their routings etc, which are also input for the process guys. It takes in general tweve months to five years, depending on the change. The new transmission has been around for five years. (Manager 2)

A number of relevant points can be highlighted from the managers' comments, regarding the conditions which they view as necessary for successful and responsive product introduction:

i) the need for top management commitment;

ii) the need for 'breaking the brick walls' by adopting an inter-functional team approach since the early conceptual stages, but making sure that the team members are really involved and that they have the appropriate level of influence and representativeness in their home functional departments; and

iii) the need for strong, high rank team leadership.

Manufacturing flexibility task and performance The managers' answers concerning the flexibility-related task of the plant are somewhat consistent, despite the fact that the answers are not exactly the same. The answers regarding the priority flexibility-related tasks which the company should focus on varied mainly between product and mix flexibility. That is understandable in a build-to-order environment, in which the distinction between mix and new product flexibility is not clear cut. The managers also seem to give priority to the range dimension of the flexibility types rather than the response dimension.

Table D.1 shows the distribution of answers.

Table D.1 Priority given by managers of Company D regarding types and dimensions of manufacturing flexibility

	Mgr1	Mgr2	Mgr3	Mgr4	Total	Total weighed
product range	4	1	2	3	4	10
product response			3		1	3
mix range	3	4		3	3	10
mix response		3			1	3
volume range			4		1	4
volume response		2	1		2	3
delivery range					0	0
delivery response					0	0

Uncertainties involved The uncertainties mentioned by the managers as the ones which represent the highest potential risk to Company D's competitiveness show a distinct pattern. All the managers, for instance, placed uncertainty with demand mix as their main concern. As a second main concern, two managers (Managers 1 and 2) pointed out overall demand volume uncertainty and two (Managers 3 and 4) pointed out uncertainty with parts and material supply. The perceived uncertainty types in Company D can be found represented in Table D.2 .

Table D.2 Importance given by managers of Company D regarding sources of uncertainty as risky for competitiveness

	Mgr1	Mgr2	Mgr3	Mgr4	Total	Total weighed
parts supply			3	3	2	6
machine breakdown		1			1	1
product introduction	1	1			2	2
demand volume	3	3	2		3	8
demand mix	4	4	4	4	4	16

It seems that the uncertainty sources which concern the managers most are those related with the demand. Three of the managers qualified Company D's suppliers as very good and reliable. The ones who pointed out parts and materials supply as an uncertainty source recognized that the uncertainty with the supply was caused by the uncertainty with the demand rather than by the suppliers themselves. The demand uncertainty is due to the broad range of customers Company D has all around the world (they export to 140 countries), with the demand therefore depending on factors such as the different government regulations, the unstable economic and political conditions in different countries and the weather conditions in different regions of the globe.

Coping with change, uncertainties and variabilities The two factors which most concern the managers at Company D in terms of uncertainty and variability are the demand uncertainty, mainly in terms of mix, and the large variety of products. Although two of the managers considered the variety of products offered by Company D as exaggerated, Manager 1 considered the variety of products the most important competitive advantage for Company D:

> Fast delivery hasn't been considered a competitive advantage. Variety, yes, this is our competitive advantage, so therefore we need to build as far as we can ultimate flexibility into the business. (Manager 1)

The most often mentioned relationships between types of uncertainty, variability and ways to deal with them are shown below, together with some pertinent quotations. The first column represents the number of managers who mentioned the relationship. The second and third columns represent the relationship itself, the second being the source of the uncertainty and the third, the capabilities or ways managers consider as worth developing to be able to cope with the uncertainty shown in the second column.

Number of managers who mentioned the relationship	Uncertainty relating to...	⇒	Best way to cope with it is by developing...
3	demand mix		rescheduling capability

> ...if we take as a given that the forecast is going to be uncertain and we still have to respond to the demand what we really need is a tool, a scheduling tool, to at least at a very early stage, identify the discrepancies between what you previously provisioned and what you now know you are going to consume... (Manager 4)

2	parts and materials supply		supplier development, partnership

> I believe the key is to have suppliers which you can trust and they can trust you, who have compatible processes, manufacturing processes ... we've got

to compress our supplier base, if you actually make the partnership, you get better commitment. They also become more responsive. They have more data accuracy about my demand more visibility to the schedule, we have greater control and support from them, so you both begin to have a great affinity and partnership, it reduces the uncertainty of the supplier. We've got to know about our process as much as we've got to know about theirs... (Manager 3)

Number of managers who mentioned the relationship	Uncertainty relating to...	⇒	Best way to cope with it is by developing...
2	demand mix		suppliers development, partnership

...what we have done as a business is to recognize that we had to change some of our logistic processes to give our suppliers more opportunities to be aware about the changes, and to respond to them ... we have to try to increase the flexibility of our suppliers helping them reduce their lead times... (Manager 1)

2	demand mix		buffer stocks

...so you've either got to convince yourself: I'm going to hold inventory strategically of certain components in places and if there is a peak, I'm going to consume it and give supplier time to react or not... (Manager 4)

2	product introduction		integration prod design/ proc design/production

...in the past we've had a very sequential way of working and therefore product changes had been very slow. In the future, we'll have simultaneous engineering and methods of working, and that will speed up the overall elapsed time. That's how we compete... (Manager 1)

2	demand mix		fast set-ups

...we, in the past, have put loads of excess labour out, we also used inventory (finished goods) to buffer that, so we made a lot of complete and unsaleable vehicles, we still do a lot of expediting, that means a lot of premium time. Now we are starting to work more on set-up times. (Manager 2)

Number of managers who mentioned the relationship	Uncertainty relating to...	⇒	Best way to cope with it is by developing...
2	demand mix		forecast sensitivity

...the demand mix is very unpredictable. We have to improve our forecasting systems. (Manager 1)

Managers in Company D also showed great concern about the large variety of the company product-line. They commented on some ways which they consider appropriate to deal with it:

Number of managers who mentioned the relationship	Variability relating to...	⇒	Best way to cope with it is by developing...
3	product variety		standardization

We can now build, theoretically, 3,640 different vehicles. I believe that what we should be doing is to simplify our products, standardize. (Manager 2)

2	product variety		supplier development/ partnership

...we are trying to tackle lead-times by trying to make our suppliers to reduce their set-up times so that we can have smaller batches and increase flexibility (Manager 1)

2	product variety		buffer stocks

I can hold level of inventory at a very low cost, which takes so much of lead-time. You are actually increasing your inventory, but very low cost inventory. That actually gives you flexibility. (Manager 3)

2	product variety		fast set-ups

Developing fast set-ups is important because we can't afford to carry on stocks of such a broad product range ... (Manager 1)

One point which is worth mentioning is the emphasis given by the management in the achievement of manufacturing flexibility through the people and infrastructure resources.

> The real fast response and day-to-day re-scheduling is made by people. We actually have to get the right balance between people and systems (Manager 4)

> ...we deal with variability by having control systems, manufacturing methods to optimize resources, regrouping the factory in cells, focusing the cells to become a flow environment... (Manager 3)

Technological resources do not seem to play a major role in the view of Company D's management as a way to achieve system flexibility. This is possibly due to the inherited inflexibilities of the machinery mentioned by Manager 2:

> We did the classic. This factory was set up for high volume, low variety. It has diversified afterwards. (Manager 2)

If that is the case, the managers would be conscious that they could not do much in terms of improving the flexibility of the system by using the technological resources, apart from attempting to reduce set-up times as much as possible.

Another point worth mentioning is the great concern placed on product variety by the majority of the managers. All of them mentioned variety as a source of complexity and argued that parts variety should be reduced although they seemed to recognize that product variety is a competitive advantage for Company D.

It seems that some of the managers view flexibility as something they have to develop in order to achieve product variety - seen as a competitive advantage - but they generally prefer to control variety and uncertainty as much as possible via parts standardization, focus, improved forecasting systems and so on, in order to reduce the need to be flexible. Generally, when asked how they dealt with variability and uncertainties, they first mentioned control-related measures (e.g. standardization) aimed at reducing the environmental changes which Company D would have to face. Then, when asked how they dealt with the effects of the changes, given that the changes had already occurred, they would mention flexibility-related measures (e.g. re-scheduling).

In terms of the uncertainty regarding demand mix, for instance, managers firstly mentioned the development of a co-operative relationship with suppliers and the development of better forecast systems aimed at reducing the uncertainty the system would have to deal with. On the other hand they pointed out that fast set-ups and buffer stocks and rescheduling capability should be used to cope with the effects of the mix demand uncertainty, when the uncertainty and the variability are taken as given.

The number of managers mentioning the ways they cope with demand mix uncertainty gives a dimension of the importance they attributed to this factor. Additionally, they emphasized the product variability, with nine mentions by the managers. They suggested standardization and co-operation with suppliers and buffer stocks and fast set-ups (flexibility-related) as ways to deal with variability. This list (except for the item standardization, exclusive for the variability, and forecasting, exclusive for the uncertainty) is similar to the one the managers suggested as being appropriate ways to deal with mix uncertainty. This indicates that in an environment like Company D's, with a broad product line and where products are built to order, the uncertainty of the mix and the variability of products are regarded by managers as calling for similar sorts of resource characteristics.

The relationship between flexibility-related competitive criteria and resource characteristics or critical success factors In general, there seems to be some consistency in the perception of the various managers interviewed about the flexibility-related competitive criteria. The four of them, for instance, ranked either product flexibility or mix flexibility as the flexibility-related priorities Company D should pursue (except for Manager 3 who also included volume range flexibility). In an environment like that of Company D which builds to order, it is understandable that mix and product flexibilities are treated, in a way, indiscriminately. The consistency was also high between the flexibility-related order winning criteria and the resource characteristics which they considered critical success factors. Table D.3 shows the two first flexibility-related priorities and the two first critical success factors according to the perception of each manager:

Table D.3 Relationship between flexibility-related competitive criteria and critical success factors, according to managers in Company D

Manager	Two first flexibility-related competitive criteria	Two first critical success factors
1	product range	design capability
	mix range	fast set-ups
2	mix range	rescheduling capability
	mix response	coordination with suppliers
3	volume range	rescheduling capability
	product response	integration design/production
4	mix response	rescheduling capability
	mix range	coordination with suppliers

The only apparent inconsistency is with the answers of Manager 3. He places a high priority on rescheduling capability and integration design/manufacturing which can not be directly related to volume range, his first priority in terms of flexibility-related manufacturing task.

For the rest of the managers, the two columns seem to show consistency between system objectives and resource characteristics, or in other words, means to achieve them.

The relationship between the uncertainties considered risky by the managers and the flexibility-related manufacturing task they regard as priorities was also found high. The reason for the consistency is possibly that the uncertainty which mostly concerns the managers is demand-related. Such uncertainty is probably caused by conditions which are intrinsic for the market Company D serves. That means that these uncertainties are also opportunities for the companies which manage to respond to and cope with such uncertainties. Therefore, the demand-related flexibility types used in the interviews are able to capture what is needed to cope with the demand-related uncertainties that are mentioned. This would possibly be different (as can be seen in the other three cases of this research work) if the managers' main concerns regarding uncertainties were in the input side or in the transformation process itself.

Summarizing, it seems that managers in Company D are able to identify the relationships between flexibility-related order winning criteria and the critical success factors which are needed to achieve high performance in them. They appear to be able to focus their attention and give priority attention to these factors.

Some conclusions of the within case D study

Managers in Company D see manufacturing flexibility as a way to cope with environmental and internal uncertainties when the causes of such uncertainties cannot be eliminated or reduced. They also consider that variability and different types of uncertainty call for different types of flexibility-related resource characteristics.

Managers' perceptions in Company D are reasonably consistent in terms of the flexibility-related manufacturing task which they should pursue. They also seem to have a consistent understanding of which ways would achieve such a manufacturing task.

Managers in Company D consider that manufacturing flexibility is generally necessary for dealing with broad product lines even when the demand is predictable. They also prefer controlling the variety of products (when such variety does not represent competitive advantage) and parts as much as possible, rather than developing the flexibility needed to deal with it.

The cross-case analysis

In the cross case analysis, the aim is to identify and analyze the differences and similarities among the particular cases. The first result found in the comparison of the case studies is that there is not a clear cut difference between the ways managers regard the management of uncertainty and variability in the Brazilian cases and the British cases. There was, for example, a major concern in all the cases regarding the supply network as a source of uncertainty which can be risky to the company's competitiveness. Other similarities are discussed below.

The similarities

The preference for uncertainty and variability control Although approaching the problem of dealing with change from a multitude of viewpoints, the interviewed managers generally showed a preference for attempting to control the sources of the uncertain and variable changes (trying to reduce their impact), rather than developing ways to respond to such changes, or in other words, developing flexibility. Not all the managers are able to discriminate clearly between restraining the occurrence and dealing with the effects of the changes but the fact is that intuitively they seem to prefer change restriction. This point is extensively discussed in chapter 4.

Supply chain uncertainties In general terms, managers in all case companies showed great concern about the uncertainties related to the supply chain as being risky to the company's competitiveness. The exception was Company C, which is the most vertically integrated of all, with more than 90 per cent of their components made in. Even in this situation, although two of Company C's managers considered the uncertainties with the government policies as the most risky, Company C's operations manager considered parts and materials supply as the most risky uncertainty factor for the company's competitiveness.

The approach for dealing with the uncertainties with supplies was also somewhat similar among the managers in all companies. Preferably, they try to increase the control over the suppliers. The ways they use for doing so vary, though. Whereas Company C tended predominantly to integrate vertically, Company A, Company B and Company D tended to try to develop long term contract and co-operation with suppliers. In the very short term and because it is impossible to get rid of the stochastic component of the supply function process, managers in all companies emphasized the need to develop skills in rescheduling the production quickly to allow the company to remain functioning by using the material which is available at the moment (when a faulty supply is identified).

The differences

The government intervention component Certainly one of the great differences between companies is the concern with the unpredictability of government policies. Such a factor was not even mentioned by any of the managers of the British case companies whereas it was mentioned by two out of three managers in Brazilian Company C as their greatest concern regarding uncertainty factors. In the other Brazilian company, Company B, all the managers mentioned the unpredictability of the government policies but they actually do not feel so vulnerable to this unpredictability since none of the managers mentioned the government actions as risky for the company's competitiveness. These differences between manager's perceptions in two companies subject to the same government actions may be due to the fact that Company C has a considerable amount of its turnover originating from exports, which can be directly affected by government actions regarding, for instance, exchange rate mechanisms. Even Company D, one of the British ones, which exports 90 per cent of its production to around 140 countries, did not mention government intervention as its concerns. The exchange rates were mentioned, though, as a source of uncertainty mainly relating to American currency, which is the reference currency for most of the countries with which Company D deals.

The emphasis given to different types of resource for achieving flexibility Company A's managers consistently considered that system flexibility is primarily achieved through the human resource whereas Company C's managers in general give less emphasis to the human aspect of system flexibility. Instead, they rely much more on managerial and information systems to achieve the levels of flexibility they need. Company D's managers, in turn, regard systems and people as the key resource types for the achievement of manufacturing flexibility. This is possibly due to the inherited inflexibility of the machinery which Company D's plant has (the company was initially set up to produce high volumes of a low variety of products). Chapter 4 discusses the roles of the different resource types in the achievement of manufacturing systems flexibility.

The arsenal of solutions for coping with uncertainty changes Managers in Brazilian Company B produced the largest arsenal with which to handle uncertainties of all the case companies. That is not surprising, for the Brazilian industrial environment is notoriously less predictable than the British one. Furthermore, the other Brazilian case company is vertically integrated at the level of 90 per cent and virtually a monopolist in its domestic market.

The concern about manager behaviour under changing circumstances It was remarkable that, in Company B managers' views were consistent in terms of their concern about the uncertainties with the middle managers' behaviour under changing circumstances. Four out of six managers interviewed in Company B mentioned that point as risky to the company's competitiveness. There was, however, no mention of that aspect by managers in any other case companies. This fact could be explained by the peculiar situation in which Company B currently finds itself. All the present line of products will die out as OEM products by the end of 1997. In the meantime, a completely new technology will be introduced: that of electronic fuel injection. The middle management probably are feeling very insecure about their jobs and demonstrating this anxiety in their attitudes as of now. They will need to learn a new technology from scratch and this effort is being undertaken already. The new technology is based primarily on electronics principles whereas the present one is based on mechanics principles.

Summary: types of uncertainty and types of flexibility-related critical success factors

The most frequently mentioned relationships between uncertainty types, variability and resource characteristics considered the most appropriate to deal with them are shown in Tables A2.1 and A2.2, respectively.

Table A2.1 Summary of the cases: the most frequently mentioned relationships between uncertainty types and flexibility-related resource characteristics during the field work

Uncertainty relating to... ⇒	Best way to cope with it is by developing...
parts and materials supply	rescheduling ability
parts and materials supply	coordination with suppliers
parts and materials supply	buffer stocks
parts and materials supply	internal machine capability
product mix	rescheduling ability
product mix	fast setups
product mix	semi finished goods stocks
product mix	ability to obtain short supplier lead times
machine breakdowns	preventive maintenance
machine breakdowns	fast corrective actions
machine breakdowns	ability to re-route production
labour absenteeism	labour multi-skills
labour absenteeism	some excess capacity of labour
product introduction	integration design / production
product introduction	ability to subcontract supply
management behaviour under changingg circumstances	training and awareness
demand	forecasting systems
labour supply	internal training
government intervention	short lead times
technology information	ability to subcontract supply
unions behaviour	close monitoring

With regard to variability:

Table A2.2 Summary of the cases: the most frequently mentioned relationships between variability and flexibility-related resource characteristics during the field work

Variability	⇒	Best way to cope is by developing...
product variety		standardization
product variety		buffer stocks
product variety		fast setups
product variety		coordination with suppliers

Ways managers cope with uncertainties and variability

The formulation of the questions made it clear that what was being asked was which ways were considered by the managers to be appropriate for dealing with their current level of uncertainties related to the various factors (since the primary aim of this research relates to the flexibility aspect). The answers therefore were not expected to include primarily the ways the managers consider appropriate for reducing the levels of uncertainty they have to deal with. However, some managers showed the preference for adopting preventive measures against the uncertainties so strongly that they frequently insisted on mentioning ways to exercise control over the uncertainty-related and variability-related changes, before mentioning ways to respond or adapt to the uncertainties and variability. This can be noticed by recalling the number of managers who mentioned, for example, preventive maintenance and coordination with suppliers to reduce the uncertainties with machine breakdowns and parts and materials supply respectively. The way the question was formulated can also explain why the number of answers regarding flexibility aspects is still higher than the number of answers regarding control, whereas it has been said in this report that in general managers have a preference for controlling the changes they would otherwise have to deal with. The relationships between different types of uncertainty and variability and the best way to deal with them, according to the managers, will be explored further below:

Parts and materials supply This is by far the uncertainty factor which appears most frequently in the mentioned relationships. The managers deal with it:

i) by developing system's rescheduling ability. Eight of the managers mentioned the system's rescheduling ability as a way to deal with it. Once a faulty supply is identified, the schedule has to be redone to allow the system to continue functioning, processing the next order which already has the necessary material available. Interestingly, all the interviewed companies except for one (company D) had rescheduling systems which were based on some key people's ability. They said that the formal systems help check availability of materials, but other factors also have to be considered such as orders priorities, subsequent bottlenecks, among others, which the formal systems can not keep up with. They also mentioned the slow responsiveness of the computerized systems for very short notice changes. (MRP II was being run once a week in the three companies which had it installed.)

ii) by developing coordination with suppliers. Six managers mentioned the need to develop closer links with suppliers in order to reduce the level of uncertainty in the interface between the customer-company and the supplier-company. They mentioned, among others, longer term contracts with reduction of the supplier base, cooperation, technical collaboration and intense information interchange as ways which reduce the uncertainties of the interface customer-supplier. The full 'partnership relationship' is generally considered by the managers as a goal, but they consistently recognized that they still had a long way to go in developing such a relationship. Meanwhile, other sort of solutions should therefore be used to cope with the effects of the current confrontation-type of relationship, which most of them still have with their present suppliers.

iii) by building up buffer stocks. The managers tended to make it clear that they considered stocks as being undesirable, in principle. Nevertheless, three of them mentioned the use of what two of them called 'strategic stocks' (the timely build up of stocks of raw materials or components when they notice that a problem with supply might be imminent). One example was Company C who starts building up buffer stocks of iron powder for sintered parts every year in October because they know it is likely that the North American winter will cause delays in the transportation of the powder, imported from the United States. (See chapter 4, for a discussion on the role of buffer stocks in the development of manufacturing flexibility.)

iv) by developing internal capability. Some managers consider that having a broad internal capability is a good way of dealing with uncertain supplies. In Company B, for instance, it is not uncommon that parts which are received below the quality specification levels are reworked in. That also happens when the suppliers are too busy to complete the part. Sometimes Company B accepts the parts semi-finished and finishes them by using its internal capability. (See chapter 4, for a discussion on the role of the resources capability in the development of manufacturing flexibility.)

Product mix Uncertainty regarding the product mix also appeared a number of times among the relationships mentioned by the managers. Fourteen managers explicitly mentioned ways they considered to be appropriate alternatives for dealing with product mix uncertainty and these ways can be put into four groups:

i) by developing rescheduling ability in order to be able to respond quickly to the changing demand mix. Interesting enough, the two companies whose managers considered their companies' ability to reschedule to be good (Company A and Company B) had the rescheduling made by a skilful scheduler, rather than by a system. On the other hand the company whose rescheduling system was almost completely automated by an MRP II system was considered by its own managers to have a poor performance in terms of rescheduling.

ii) by developing fast set-ups. With fast set-ups, some managers argued, the cost and times for changeover are reduced, allowing for quicker response to the changes. All the case companies had programs running on set-up reduction. Set-up reduction was consistently one of the first aspects to be mentioned by managers when talking about developing flexibility in general. In a first approach, before analyzing types and dimension of flexibility, managers seem to associate flexibility very closely with fast set-ups.

iii) by having stocks of semi-finished goods. Stock piling components and parts and assembling them or configuring them to order was also mentioned as a way to respond quickly to demand mix changes. Another interesting approach for reducing the response time to mix changes is used by Company A. As the engines they make have a number of common parts and features, they have rearranged the assembly operations sequence along the track to assemble all the common features first and the special features, those which actually differentiate the engines, later. In doing so, they have managed to reduce the time elapsed to changeover products in the assembly line from seven hours to two hours on average.

iv) by developing the ability to get short supply lead-times. According to the managers, this can be achieved either by efficient procurement or by coordination with suppliers. Having short supply lead-times, companies would be able to respond better to unexpected changes in their product mix.

Machine breakdowns This was mentioned ten times by various managers. The two approaches in dealing with unexpected changes are quite clear cut in this aspect of the case studies:

i) by developing preventive maintenance, which is another example of the preference of the managers for controlling or restraining the occurrence of the changes. Five of them mentioned that the best way to deal with breakdowns is by not allowing them to happen.

ii) by developing the ability to take fast corrective actions. Once the breakdown occurs, acknowledging the occurrence and mobilizing the right resources to have the machine up and running again is what fast corrective action is about, according to the managers who mentioned it. Company B, for instance, keeps a separate budget, controlled by the production manager (for the urgent replacement parts purchase not to have to pass through the purchasing department) and even a car specially dedicated to fetching the replacement parts in case of critical breakdowns. This scheme is kept in parallel with a preventive maintenance program also being worked on. (See chapter 4 for further discussion on this issue).

iii) by developing rerouting ability aimed at bypassing the broken machine. To be able to do this quickly, Company B has recently done a study on 'what machine can perform what part' and displayed the result on a big board in each manufacturing unit. This way, the foreman can quickly redirect all the critical parts in the broken machine queue to other machines which can perform the operation the part was queuing for.

Labour absenteeism This was mentioned nine times. A consistent level of absenteeism was found in all companies - all of them varying from 3 to 6 per cent. The managers mentioned two basic ways of dealing with it:

i) by having some excess of labour capacity. All the companies keep some excess workers for absenteeism cover, ranging from 3 to 6 per cent. However, excess capacity is not enough for companies to be totally covered against absenteeism. Company C's assembly line, for instance, needs 50 people to be run, but they need the right 50 people. One can even be sure that the next morning 50 people will be ready to work in the line and still one cannot tell whether he is going to have the right set of skills to run the line. The solution mentioned by several managers is to develop multiskilled workers.

ii) by developing multi skills. If some of the workers at a production unit are trained to perform a number of the unit's tasks, it is more likely that within the 50 people of the example above there will be the right skills to run the line.

New product introduction Uncertainties with new product introduction and product changes (regarding launch dates, specifications and so on) were mentioned by six managers. The ways they consider appropriate for dealing with them are:

i) by developing inter-function integration between product design and development, process development and production: to make sure that the products are designed to manufacture right the first time. A number of aspects of this integration were mentioned by one of Company A's managers, describing a recent and very successful introduction of a new engine to equip a new car (the elapsed time between the initial conceptual ideas about the car and the first car to be assembled regularly was reduced to three years, a record for the company): a multi function team approach, early involvement of direct workers in the design and prototyping phases, early involvement of suppliers and delegation to expert firms of the task of designing and developing the parts, a reduction of the number of suppliers and a tendency to establish long term contracts with them.

ii) by development of the ability to subcontract supply. Company A's and Company C's managers mentioned the ability to subcontract supply as a way of dealing better with the uncertainties regarding new product introduction. However, they mentioned subcontract supply for two different reasons. Company C has a very high occupation rate, bottlenecked, in the words of one of its managers, not only in terms of production but also in the product development function. One of Company C's managers mentioned that there was a queue of six months to get a new die made, because of the company's machining shop overutilization. The manager thinks that if the company subcontracted external companies to make the dies they would be able to respond much better to the demand for new products. The Company A manager who mentioned supply subcontract, on the other hand was referring to a different type of supply. According to him, Company A is following a trend in the automotive industry: the car assemblers would be delegating the task of designing and developing the parts to expert firms. Company A, for instance, had always designed its own diesel engines. With the new laws and regulations regarding emissions, the technology that is entailed has evolved very quickly in recent years and hence Company A preferred to buy in the design of its new diesel engine from an expert firm in Italy. (Chapter 4 discusses the role of subcontracting and a number of other actions in managing unplanned change.)

Management behaviour relating to change All four managers who mentioned management behaviour were Company B's managers. What probably made them mention this aspect was the peculiar situation of the company at the time of the interviews. In six years time they expected the whole of the current line of products to die as OEM products. A new technology (electronic fuel injection) and the introduction of a totally new product line was being planned, which, according to the managers, had caused mixed feelings among the middle managers. On the one hand, they were motivated by the challenge, but at the same time they were somewhat anxious and insecure because of the unknown. For the managers interviewed, the way they were dealing with this uncertainty is by emphasizing manager training and improving their awareness about the changes to come.

Demand Uncertainties about demand were mentioned by three managers who associated it with the need for the development of better forecasting systems, to reduce such uncertainty.

Labour supply This was mentioned by two of Company C managers who were, at the time of the interviews, facing a problem with finding the right skills to man a new factory which Company C had opened in a remote country side region. The way to deal with this uncertainty, according to the managers, is to intensify the training done within the company. This way, they can recruit people who are not qualified and provide them with the right skills.

Government intervention This was mentioned by two out of three Company C managers as the most risky factor for their competitiveness. This concern is probably due to the dependence of Company C on export markets to support its strategy, which in turn is dependent on government policies regarding exchange rates. Unexpected changes in the exchange rates can make what seemed a good deal at the sale become a bad one at the time the payment is made. In the managers' view, the only way out is to reduce all the cycle times involved in the production to reduce the company's product lead-times. With shorter lead-times, according to the managers, the company is less vulnerable to such uncertainties.

Unions behaviour This was mentioned by two managers in Company C, probably because they are located in a region where the unions are very powerful. They saw close monitoring of union behaviour as the only way to avoid an unexpected strike, for instance.

Product variety This was mentioned nine times by the managers. They approached the variety issue in several ways:

i) by developing standardization: managers mentioned standardization of parts as well as standardization of products. They emphasized that the best way of doing this is by designing out excess variety and actually having different parts among products, exclusively for the parts which represent the difference between products which the customer actually values.

ii) by developing coordination with the suppliers, in order to help them shorten lead-times, set up times and therefore to reduce their lot sizes. This should make them more able to respond to a higher variety of parts.

iii) by having strategic buffer stocks of some parts with low level of value added, aiming at reducing their perceived lead-times.

iv) by having fast set-ups: the more quickly switchable the resources are, the greater the variety of products the system can provide within a certain period.

Some conclusions of the case studies

The managers considered flexibility as one of the ways to deal with change in organizations, mainly when the change perceived by the organization cannot be eliminated by restraining its occurrence.

The managers did not always discriminate explicitly between control and flexibility and did not always have a clear view of what should be the most appropriate way to deal with the different types and dimension of change. However, they were able to mention a number of ways they actually use in order to reduce the levels of uncertainty and variability they have to manage and also a number of ways they see as alternatives for reacting to the changes they did not control for some reason.

The managers understand that different types of change should be dealt which by developing different types of resources. However, they in general do not seem to have a consistent model to help them make decisions in that regard, which causes anxiety and sometimes frustration to a number of them.

The managers who are more aware about manufacturing flexibility see flexibility as a reserve, something which should be planned for, developed, maintained and considered as a valuable asset of the organization. However, this view is intuitive and the managers were not able to explain it or analyze it in more depth.

The managers who face the most uncertain situation with regard to supply (erratic or uncertain supplies, for example) and process (unreliable machines, for example) develop specific characteristics of the set of their structural and infra-structural resources in order to increase the reliability of the manufacturing system. This includes not only procedures which aim at increasing the reliability of the individual resources, such as preventive maintenance, but also procedures which involve the set of inter-acting resources.

The managers generally found Slack's (1988) classification of flexibility in types and dimensions understandable and useful, at least for describing the flexibility aspects which are related to the output of the manufacturing systems.

Notes

1. Manufacturing Resources Planning System is a computer based method for planning manufacturing resources based on calculating the resource requirements so that the orders meet the due dates and on checking with the available capacity. For a complete discussion on MRP systems, see Vollmann et al (1988).
2. This is a very industrialized region in Sao Paulo, where most of the automotive industry plants are located. It is a place where the unions are very powerful.

Bibliography

Adler, P.A. (1987), 'Managing Flexible Automation', Working Paper. Dept. of Industrial Engineering and Engineering Management. Stanford University.

Atkinson, J. (1984), 'Manpower Strategies for Flexible Organisations', *Personal Management*, August, pp. 28-31.

Benne, K.D. (1961), *Changes in Institutions and the Role of the Change Agent*, Irwin, Homewood.

Bessant, J. and Haywood, B. (1986), 'Introducing FMS', *Production Engineer*, April.

Blackburn, J. and Millen, R. (1986), 'Perspectives on Flexibility in Manufacturing: Hardware vs. Software' in *Modeling and Design of FMS*, Editor Kusiak, A., Elsevier, Amsterdam.

Brill, P.H. and Mandelbaum, M. (1989), 'On the Measures of Flexibility in Manufacturing Systems', *International Journal of Production Research*, Vol. 27, No. 5, pp. 746-56.

Bryman, A. (1989), *Research Methods and Organization Studies*, Unwin Hyman, London.

Browne, J. et al. (1984), 'Classification of Flexible Manufacturing Systems', *The FMS Magazine*, Vol. 2, No. 2, pp. 114-17.

Buffa, E.S. (1984), *Meeting the Competitive Challenge*, Irwin, Illinois.

Buzacott, J.A. (1982), 'The Fundamental Principles of Flexibility in Manufacturing Systems', *Proceedings of the 1st International Conference on Flexible Manufacturing Systems*, Brighton, pp. 13-22.

Carter, M.F. (1986), 'Designing Flexibility Into Automated Manufacturing Systems', *Proceedings of the 2nd ORSA/TIMS Conference on FMS*, Editors Stecke, K. and Suri, R., Elsevier, Amsterdam.

Chambers, S. (1990), 'Flexibility in the Context of Manufacturing Strategy', Paper presented in the 5th International Conference of the Operations Management Association UK, University of Warwick, June.

Chandra, P. and Tombak, M. (1990), 'Models for the Evaluation of Manufacturing Flexibility', Working Paper No. 90/61/TM, Insead, Fontainebleau.

Corrêa, H.L. and Slack, N.D.C. (1991), 'The Flexibility of Three Manufacturing Planning and Control Systems', *Proceedings of the OMA - UK Sixth International Conference*, Springer-Verlag, London, pp. 57-63.

Corrêa, H.L. (1992), *The Links Between Uncertainty, Variability of Outputs and Flexibility in Manufacturing Systems,* Ph.D. Dissertation, University of Warwick, Coventry, United Kingdom.

Corrêa, H.L. and Gianesi, I.G.N. (1992), 'Dynamic Manufacturing Strategy Development for Proactive Manufacturing in Brazil' in *Crossing Borders in Manufacturing and Service,* Editors Hollier, R.H. et al., North-Holland, Manchester, pp. 31-6.

Cummings, T.G. and Huse, E.F. (1989), *Organization Development and Change*, West Publishing Company, St. Paul.

Dawes, R.M. (1975), *Fundamentals of Attitude Measurement*, John Wiley and Sons, New York.

De Meyer, A. (1986), 'Flexibility: The New Competitive Battle', Working Paper No. 86/31, Insead, Fontainebleau.

Dooner, M. and De Silva, A. (1990), 'Conceptual Modeling to Evaluate the Flexibility Characteristics of Manufacturing Cell Designs', *Proceedings of the 28th Matador Conference*, UMIST, Manchester.

Downey,H. et al. (1975), 'Environmental Uncertainty: The Construction and Its Application', *Administrative Science Quarterly*, Vol. 20, pp. 613-29.

Downey, H. and Slocum, J.W. (1975), 'Uncertainty: Measures, Reach and Sources of Variation', *Academy of Management Journal*, Vol. 18, No. 3, pp. 562-78.

Duncan, R.B. (1972), 'Characteristics of Organisational Environments and Perceived Environmental Uncertainty', *Administrative Science Quarterly*, Vol. 17, pp. 313-27.

Eisenhardt, K.M. (1988), 'Building Theory from Case Study Research', Working Paper, Department of Industrial Engineering and Engineering Management, Stanford University.

Ferdows, K. and Lindberg, P. (1986), 'FMS as Indicator of Manufacturing Strategy', Working Paper No. 86/42, Insead, Fontainebleau.

Ferdows, K. and Skinner, W. (1986), 'Manufacturing in a New Perspective', Working Paper No. 86/41, Insead, Fontainebleau.

Fine, C.H. and Hax, A.C. (1985), 'Manufacturing Strategy: A Methodology and an Illustration', Working Paper, Sloane School of Management, MIT, Cambridge, Massachusetts.

Frazelle, E.H. (1986), 'Flexibility: A Strategy Response in Changing Times', *Industrial Engineering*, March, pp. 17-20.

Gerwin, D. (1982), 'Do's and Don't's of Computerized Manufacturing', *Harvard Business Review*, Mar-Apr.

Gerwin, D. and Tarondeau, J.C. (1982), 'Case Studies of Computer Integrated Manufacturing Systems: A View of Uncertainty and Innovation Processes', *Journal of Operations Management*, Vol. 2, No. 2.

Gerwin, D. (1986), 'An Agenda for Research on the Flexibility of Manufacturing Processes', *International Journal of Operations and Production Management*, Vol. 7, No. 1, pp. 38-49.

Gerwin, D. and Tarondeau, J.C. (1989), 'International Comparisons of Manufacturing Flexibility' in Editor Ferdows, K., *Managing International Manufacturing*, North-Holland, Amsterdam.

Gifford, W.E. et al. (1979), 'Message Characteristics and Perceptions of Uncertainty by Organizational Decision Makers', *Academy of Management Journal*, Vol. 22, No. 3, pp. 458-81.

Goldhar, J.D. and Jelinek, M. (1983), 'Plan for Economies of Scope', *Harvard Business Review*, Nov-Dec.

Goldhar, A.T.; Hassan, M.Z. and Talaysum, A.T. (1987), 'Uncertainty Reduction Through Flexible Manufacturing', *IEEE Transactions on Engineering Management*, Vol. EM-34, No. 2.

Goldratt, E.M.(1988), 'Computerized Shop Floor Scheduling', *International Journal of Production Research*, Vol. 26, No. 3, pp. 443-55.

Gregory, M.J. and Platts, M.J. (1990), 'A Manufacturing Audit Approach to Strategy Formulation', *Proceedings of the 5th International Conference of the Operations Management Association U.K.*, University of Warwick, Vol. 2.

Gupta, D. and Buzacott, J.A. (1986), 'Notions of Flexibility in Manufacturing Systems', Working Paper, Department of Management Sciences, University of Waterloo.

Gupta, Y.P. and Goyal, S. (1989), 'Flexibility of Manufacturing Systems: Concepts and Measurements', *European Journal of Operational Research,* Vol. 43, pp. 119-35.

Hayes, R.H. and Wheelwright, S.C. (1984), *Restoring Our Competitive Edge*, John Wiley and Sons, New York.

Hayes, R.H. et al. (1988), *Dynamic Manufacturing*, The Free Press, New York.

Hill, T. (1985), *Manufacturing Strategy*, MacMillan, London.

Hill, T. (1993), *Manufacturing Strategy*, second edition, MacMillan, London.

Huff, A.S. (1978), 'Consensual Uncertainty', *Academy of Management Review*, July, pp. 651-54.

Jaikumar, R. (1986), 'Postindustrial Manufacturing', *Harvard Business Review*, Nov-Dec.

Johnson, H.T. and Kaplan, R.S. (1987), *Relevance Lost - The Rise and Fall of Management Accounting*, Harvard Business School Press, Boston, Massachusetts.

Kaplan, R.S. (1984), 'Yesterday's Accounting Undermines Production', *Harvard Business Review*, Jul-Aug.
Kotler, P. (1991), *Marketing Management - Analysis, Planning, Implementation and Control*, seventh edition, Prentice-Hall, Englewood Cliffs, New Jersey.
Kumar, V. (1987), 'Entropic Measures of Manufacturing Flexibility', *International Journal of Production Research*, Vol. 25, No. 7, pp. 957-66.
Lawrence, P.R. and Lorsch, J.W. (1969), *Organization and Environment*, Irwin, Homewood.
Lawrence, P.R. at al. (1976), *Organizational Behaviour and Administration*, third edition, Irwin, Homewood.
Leong, G.K., et al. (1990), 'Research in The Process and Content of Manufacturing Strategy', *OMEGA, International Journal of Management Sciences*, Vol. 18, No. 2, pp. 109-22.
Likert, R. (1967), *The Human Organization: Its Management and Value*, McGraw-Hill, New York.
Lim, S.H. (1987), 'Flexible Manufacturing Systems and Manufacturing Flexibility in the United Kingdom', *International Journal of Operations and Production Management*, Vol. 7, No. 6, pp. 44-54.
Luce, R.D. and Raiffa, H. (1957), *Games and Decisions*, John Wiley and Sons, New York.
Magee, B. (1990), *Popper*, 15th impression, Fontana Press, London.
Mandelbaum, M. (1978), *Flexibility in Decision Making: An Exploration and Unification*, PhD Dissertation, Department of IE, University of Toronto, Canada.
Miles, M.B. (1979), 'Qualitative Data as an Attractive Nuisance: The Problem of Analysis', *Administrative Science Quarterly*, Vol. 24, pp. 590-601.
Miller, S.S. (1988) *Competitive Manufacturing - Using production as a Management Tool*, Van Nostrand Reinhold Company. New York.
Mintzberg, H. (1979), 'An Emerging Strategy of "Direct" Research', *Administrative Science Quarterly*, Vol. 24, pp. 580-9.
Morgan, G. and Smircich, L. (1980), 'The Case for Qualitative Research', *Academy of Management Review*, Vol. 5, No. 4, pp. 491-500.
Muramatsu, R. et al. (1985), 'Some Ways to Increase Flexibility in Manufacturing Systems', *International Journal of Production Research*, Vol. 23, No. 4,. pp. 691-703.
Neal, P. and Leonard, R. (1982), 'The Measurement of Variation Reduction', *International Journal of Production Research*, Vol. 20, No. 6, pp. 689-99.
Perrow, C. (1967), 'A Framework for the Comparative Analysis of Organisations', *American Sociological Review*, Vol. 32, pp. 194-208.

Pettigrew, A. (1988), 'Longitudinal Field Research on Change: Theory and Practice', Paper presented at the National Foundation Conference on Longitudinal Research Methods in Organizations, Austin, Texas.

Platts, K.W. and Gregory, M.J. (1990), 'A Manufacturing Audit Approach to Strategy Formulation', *Proceedings of the 5th International Conference of the Operations Management Association U.K.*, University of Warwick, pp. 636-54.

Poe, R. (1987), 'Inflexible Manufacturing', *Datamation*, June, pp. 63-66.

Schmenner, R. W (1990), *Production/Operations Management - Concepts and Situations,* Macmillan, New York.

Schonberger, R.J. (1982), *Japanese Manufacturing Techniques*, The Free Press, New York.

Schonberger, R.J. (1986), *World Class Manufacturing*, The Free Press, New York.

Schonberger, R.J. (1990), *Building a Chain of Customers*, Hutchinson Business Books, London.

Secord, P. and Backman, C. (1964), *Social Psychology*, McGraw Hill, New York.

Shingo, S. (1985), *A Revolution in Manufacturing: The SMED System*, Productivity Press, Stanford, Massachusetts.

Skinner, W. (1969), 'Manufacturing - Missing Link in Corporate Strategy', *Harvard Business Review*, May-Jun.

Skinner, W. (1974), 'The Focused Factory', *Harvard Business Review,* May-Jun.

Skinner, W. (1985), *Manufacturing: The Formidable Competitive Weapon*, John Wiley and Sons, New York.

Slack, N.D.C. (1983), 'Flexibility as a Manufacturing Objective', *International Journal of Production and Operations Management*, Vol. 3, No. 3, pp. 4-13.

Slack, N.D.C. (1987), 'Manufacturing Systems Flexibility: Ten Empirical Observations', Working Paper No. MRP87/9, Templeton College, Oxford.

Slack, N.D.C. (1988), 'Manufacturing Systems Flexibility - An Assessment Procedure', *Systems*, Vol. 1, No. 1, February.

Slack, N.D.C. (1989), 'Focus on Flexibility' in *International Handbook of Production and Operations Management,* Editor Wild, R., Cassell, London, pp. 50-73.

Slack, N.D.C. (1989a), 'Managing the Manufacturing Function', Brunel University Handout, Uxbridge.

Slack, N.D.C. (1990), 'Managing Operations: The Operations Advantage', in *The Directors Handbook*, Director's Books, London.

Slack, N.D.C. (1990a), 'The Virtues of Versatility', *Proceedings of the BAM Conference*, Glasgow.

Slack, N.D.C. (1991) *The Manufacturing Advantage*, Mercury, London.

Slack, N.D.C. and Corrêa, H.L. (1992), 'The Flexibility of Push and Pull', *International Journal of Operations and Production Management*, Vol. 12, No. 4, pp.82-92.

Stalk, Jr., G. and Hout, T.M. (1990), *Competing Against Time*, The Free Press, New York.

Stecke, K.E. and Raman, N. (1986), 'Production Flexibilities and Their Impact on Manufacturing Strategy', Working Paper No. 484, Graduate School of Business Administration, University of Michigan.

Swamidass, P.M. (1985), 'Manufacturing Flexibility: Strategic Issues', Discussion Paper 305, Graduate School of Business, Indiana University, Indiana.

Swamidass, P.M. and Newell, W.T. (1987), 'Manufacturing Strategy, Environmental Uncertainty and Performance: A Path Analytical Model', *Management Science*, Vol. 33, No. 4.

Thompson, J.D. (1967), *Organizations in Action*, McGraw-Hill, New York.

Tidd, J. (1991), *Flexible Manufacturing Technologies and International Competitiveness*, Pinter Publishers, London.

Tosi, H. et al. (1973), 'On the Measurement of the Environment: An Assessment of the Lawrence and Lorsch Environmental Uncertainty Subscale', *Administrative Science Quarterly*, Vol. 18, pp. 27-36.

Tverski, A. and Kahneman, D. (1989), 'Judgement under Uncertainty: Heuristics and Biases' in Editor Diamond, P., *Uncertainty in Economics: Readings and Exercises*, Academic Press, New York.

Voss, C.A. et al. (1985), *Operations Management in the Service Industries and the Public Sector*, John Wiley and Sons, London.

Voss, C.A. (1986), 'Implementing Manufacturing Technology: A Manufacturing Strategy Approach', *International Journal of Operations and Production Management*, Vol. 6, No. 4, pp. 17-26.

Wieland, G.F. and Ullrich, R.A. (1976), *Organizations: Behaviour, Design and Change*, Irwin, Homewood.

Wild, R. (1980), *Operations Management: A Policy Framework*, Pergamon, Oxford.

Wild, R. (1989), *Production and Operations Management*, fourth edition, Cassell, London.

Womack, J.P et al. (1990), *The Machine That Changed the World*, Rawson Associates, New York.

Yin, R.K. (1989), *Case Study Research - Design and Methods*, Revised edition, Sage Publications, Newbury Park, California.

Zelenovic, D.M. (1982), 'Flexibility: A Condition for Effective Production Systems', *International Journal of Production Research*, Vol. 20, No. 3, pp. 319-37.